Lecture Notes in Mathematics

Volume 2389

This series reports on new developments in all areas of mathematics and their applications - quickly, informally and at a high level. Mathematical texts analysing new developments in modelling and numerical simulation are welcome. The type of material considered for publication includes:

1. Research monographs
2. Lectures on a new field or presentations of a new angle in a classical field
3. Summer schools and intensive courses on topics of current research.

Texts which are out of print but still in demand may also be considered if they fall within these categories. The timeliness of a manuscript is sometimes more important than its form, which may be preliminary or tentative. Please visit the LNM Editorial Policy (https://drive.google.com/file/d/1MOg4TbwOSokRnFJ3ZR3ciEeK s9hOnNX_/view?usp=sharing)

Titles from this series are indexed by Scopus, Web of Science, Mathematical Reviews, and zbMATH.

Ercília Sousa

Finite Difference Methods for Fractional Diffusion Equations

One-Dimensional Time-Dependent Problems

Ercília Sousa
Depto. de Matemática
Universidade de Coimbra
Coimbra, Portugal

ISSN 0075-8434   ISSN 1617-9692   (electronic)
Lecture Notes in Mathematics
ISBN 978-3-032-11221-7   ISBN 978-3-032-11222-4   (eBook)
https://doi.org/10.1007/978-3-032-11222-4

Mathematics Subject Classification: 35R11, 65M12, 65M80, 65M06, 65M15

This Springer imprint is published by the registered company Springer Nature Switzerland AG
The registered company address is: Gewerbestrasse 11, 6330 Cham, Switzerland

To Luís
who may one day unlock these pages

Preface

The integro-differential equations studied in this book bring together two fundamental concepts: fractional derivatives and diffusion. The theoretical foundations
and practical applications of fractional derivatives over the past several decades
have experienced rapid and continuous development, particularly in connection with
modeling anomalous transport and memory effects in complex systems. Diffusion,
on the other hand, remains a cornerstone of mathematical physics, describing a
wide range of natural and engineering processes. The combination of these two
ideas gives rise to space-fractional diffusion equations, which provide a powerful
framework for capturing nonlocal and heterogeneous phenomena that cannot be
adequately represented by classical models. In this book, our primary objective
is to study finite difference approximations of one-dimensional time-dependent
space-fractional diffusion equations, with a view toward understanding both their
theoretical properties and their computational implementation.

Leibniz, Euler, Laplace, Lacroix, and Fourier mentioned derivatives of arbitrary
order related to the question of the extension of meaning of a derivative of non-
integer order [117]. But it is believed that it was Niels Henrik Abel in 1823 that first
arrived at the definitions of fractional integral and derivative [110, 117] as a result
of trying to find an analytical solution of an integral equation that formulates the
tautochrone problem [2]. The now known as the Abel integral equation can be found
in many textbooks, including in the next chapter of this book, since through this
equation we can explain the rationale behind the definition of fractional derivative.
It is only in 1974 that we can find a published textbook by Oldham and Spanier,
completely dedicated to this topic [104] and apparently in the same year the first
conference on the topic was held. Its proceedings were edited by Bertram Ross and
published by the Lecture Notes in Mathematics [116].

By the end of the last century, we could already find many works in the field of
physics addressing equations with non-integer order derivatives. In 2000, Metzler
and Klafter presented a very comprehensive and complete review about fractional
differential equations related to anomalous diffusion [98]. The authors mention in
the abstract that their work demonstrates that *"fractional equations have come of
age as a complementary tool in the description of anomalous transport processes."*

The appearance of fractional differential equations became very appealing due to their proximity to the analogous standard equations, including the standard diffusion problems.

Diffusion is the process by which matter is transported from one part of a system to another as a result of random molecular motions. The mathematical description of diffusion has a long history with many different formulations based on conservation of mass and constitutive laws. A fundamental result common to the classical approaches is that the mean square displacement of a diffusing particle scales linearly with time. However, there have been numerous experimental measurements where this phenomenon is not observed and the diffusion particle scales as a fractional order power law in time. To deal with these observations, in recent years a great progress has been achieved in extending the different models for diffusion. Some of these changes have consisted in the incorporation of fractional derivatives in the diffusion models. They seem to be a powerful instrument for the description of memory and hereditary properties of different substances. Therefore, they have been included in equations that appear in many subjects.

We mention few of the subjects that have models related to fractional differential equations: to simulate transport processes in contaminant hydrology [14, 67, 176]; to describe human traveling [19]; to describe animal movement [56]; to model magnetic resonance imaging through a modified Bloch-Torrey equation [91, 177]; to simulate the behavior of financial markets [111]; to help describing turbulent fluid flow [10, 34, 124, 133]; to improve existent tools in image processing [11, 30, 114, 170].

The diffusive models we will be discussing here can be derived within the framework of Lévy flights, which name is attributed to Mandelbrot [94, page.289]. The Lévy flights are Markovian processes that include jumps and long-distance interactions, resulting in anomalous diffusion. These processes are related to the Lévy alpha-stable distributions, which are the fundamental solutions of the fractional advection diffusion equations. The fractional derivatives, which appear in the context of Lévy flights, are integro-differential operators describing a nonlocal process. Its approximation is a very challenging task, since they are defined as the convolution of a singular kernel with the function which derivative we want to determine.

A fascinating detail about this research area is that there are still many open questions, which have a simple formulation, but are still difficult to answer, such as the ones related to the presence of boundary conditions. Therefore, we include the insufficiently discussed subject of the presence of boundaries for fractional differential equations related to Lévy flights. We present some of the theoretical background and tools that have been used to tackle some of the main questions.

The deterministic equations that describe the anomalous diffusion related to Lévy flights are obtained through the Fourier transform. The Fourier transform is defined in an open domain, not a bounded domain, and therefore it is easy to understand why in the presence of boundaries we have several open questions either from the physical point of view or from the mathematical point of view. Research on the subject of including boundary conditions into problems based on Lévy flights has been emerging recently. From the mathematical point of view, where the well-posedness of the problem is the main concern, some of the approaches do not have a satisfactory physical interpretation. On the other hand, some of the physical explanations do not have a clear mathematical translation. It is therefore a good idea to pay more attention to the physics and mathematics interplay. Additionally and unsurprisingly most of the resulting partial integro-differential equations cannot be solved in closed form. In the few situations where analytical solutions can be found, the expression is so cumbersome, that very little or no information can be taken from its analytical form. Consequently, the access to the numerical solutions of such problems is of great importance and may shed some light into many open questions.

The subject of fractional differential equations is too rich to be discussed solely in one course. It is our opinion that the problems with only fractional derivatives in space related to the diffusive phenomena deserves a graduate course and therefore a book. We have decided to focus in the anomalous diffusion described by spatial fractional operators, not only for its importance alone, but also because it is a natural way to introduce the subject of finite difference methods to fractional differential equations, in the same way that, for classical equations, parabolic equations are used to introduce finite difference methods and elliptic equations to introduce finite element methods.

We give an overview of the content of the next chapters. In the first chapter, we present some fundamental properties about fractional derivatives including how the definition arrives naturally with the help of the Abel integral equation. This will remind us of the Fundamental Theorem of Calculus for standard derivatives. In the second chapter, we describe the relation between Lévy flights and fractional advection diffusion equations through the Fourier transform. The main purpose of this book is to teach how to derive numerical solutions for this type of models, and this is addressed in the rest of the book. The third chapter gives an overview of consistency, stability, and convergence of finite difference methods for fractional differential equations. In the fourth chapter, numerical methods are presented for problems defined in the whole real line. Then, we include two chapters about the presence of boundaries for equations related to Lévy flights. The chapter of absorbing boundary conditions is well developed, since in literature there is already an intensive discussion. The last chapter, on reflecting boundary conditions, can be described as a more exploratory chapter. We finish with an outline of the subjects that are the focus of ongoing research.

We assume the reader is familiar with standard calculus and the basic concepts of linear algebra and numerical analysis. The lectures notes can be used to teach a course during a semester at a graduate level or at summer schools. It can also be used to teach part of a graduate course in finite difference methods, where the first part can be about standard diffusion equations and the second part can be about fractional diffusion equations. In this case, Chaps. 1, 2, and 4 can be used, if there is no time to discuss the presence of boundaries.

The book is structured such that each chapter is self-contained. Therefore, if readers are already acquainted with the material covered in the initial chapters, they can navigate to the latter chapters of the book. At the end of each chapter suitably designed exercises are included to help readers to deepen the understanding of the subjects. Each chapter has approximately ten exercises, along with their respective solutions.

Coimbra, Portugal   Ercília Sousa

Acknowledgments I would like to begin by expressing my gratitude to the Department of Mathematics at the University of Coimbra for granting me a sabbatical year in 2022–2023, which provided me with the opportunity to start writing the book I had been envisioning for quite some time.

I am also deeply thankful to my colleagues and friends within my department, as well as those around the world whom I have had the pleasure of meeting at conferences and during academic visits, for the many insightful discussions we have shared on related topics. Although sometimes not directly tied to this book, those exchanges have greatly enriched my intuitive understanding of the concepts and helped me find clearer ways to articulate certain ideas. In this regard, I am also grateful to the PhD students to whom I have taught portions of the chapters. Their reactions in the lecture room helped me refine and improve various parts of the book.

On a more personal note, I would like to thank Alexandre and Lurdes, who subtly nourish the enthusiasm with which I approach life and research.

Competing Interests The author has no competing interests to declare that are relevant to the content of this manuscript.

Contents

Chapter 1
Fractional Derivatives: Fundamental Concepts

In this introductory chapter, we present definitions and properties regarding fractional derivatives and integrals. The main focus concerns the Riemann-Liouville fractional derivative and its properties when the function is defined on a finite interval. In the end, we present results that link this fractional derivative with the also well known Grünwald-Letnikov fractional derivative and the Caputo fractional derivative.

1.1 Riemann-Liouville Fractional Integrals

The definitions and properties of fractional integrals and derivatives that are going to be presented can be found in a number of interesting books, such as, [75, 104, 109, 119].

We start to describe how we arrive at the definition of fractional integral and then how this leads naturally to the definition of fractional derivative.

Suppose that u is continuous and integrable in every finite interval (a, x). The function u may have an integrable singularity of order $r < 1$ at the point a, that is,

$$\lim_{\tau \to a} (\tau - a)^r u(\tau)$$

exists. Then, the integral

$$I_{a+}u(x) = \int_a^x u(\tau)d\tau$$

has a finite value. In particular

$$\lim_{x \to a} I_{a+}u(x) = 0.$$

© The Author(s), under exclusive license to Springer Nature Switzerland AG 2026
E. Sousa, *Finite Difference Methods for Fractional Diffusion Equations*, Lecture Notes in Mathematics 2389, https://doi.org/10.1007/978-3-032-11222-4_1

Consider the integration of the previous integral, defined as $I_{a+}^2 u(x)$,

$$I_{a+}^2 u(x) := \int_a^x dx_1 \int_a^{x_1} u(\tau)d\tau.$$

It is easy to see that, by doing a change of variables, we can write

$$I_{a+}^2 u(x) = \int_a^x u(\tau)d\tau \int_\tau^x dx_1$$

$$= \int_a^x u(\tau)(x - \tau)d\tau.$$

Integrate again and denote it by $I_{a+}^3 u(x)$, that is,

$$I_{a+}^3 u(x) := \int_a^x dx_1 \int_a^{x_1} dx_2 \int_a^{x_2} u(\tau)d\tau.$$

As previously we can write

$$I_{a+}^3 u(x) = \int_a^x dx_1 \int_a^{x_1} u(\tau)d\tau \int_\tau^{x_1} dx_2$$

$$= \int_a^x dx_1 \int_a^{x_1} u(\tau)(x_1 - \tau)d\tau.$$

Then, it follows

$$I_{a+}^3 u(x) = \int_a^x u(\tau)d\tau \int_\tau^x (x_1 - \tau)dx_1$$

$$= \frac{1}{2} \int_a^x u(\tau)(x - \tau)^2 d\tau.$$

For the general case,

$$I_{a+}^n u(x) := \int_a^x dx_{n-1} \int_a^{x_{n-1}} dx_{n-2} \ldots \int_a^{x_3} dx_2 \int_a^{x_2} dx_1 \int_a^{x_1} u(\tau)d\tau.$$

We can prove by induction that we have the Cauchy formula

$$I_{a+}^n u(x) = \frac{1}{(n-1)!} \int_a^x u(t)(x - t)^{n-1} dt. \tag{1.1}$$

Recall that the Gamma function $\Gamma(z)$ is defined by the integral

$$\Gamma(z) = \int_0^\infty e^{-t} t^{z-1} dt,$$

which converges in the right half of the complex plane $Re(z) > 0$ and it satisfies the property $\Gamma(z + 1) = z\Gamma(z)$. Therefore, since the Gamma function coincides with the factorial function for integers, that is, $\Gamma(n) = (n - 1)!$, the next definition arrives naturally if in the integral on the right hand side of (1.1) we substitute the factorial by the Gamma function and n by α.

Before introducing the fractional integral, we first recall the spaces of integrable functions. For a real number $p \geq 1$ and an open set $\Omega \subset \mathbb{R}$, we denote by $L_p(\Omega)$ the space of all Lebesgue-measurable functions u defined on Ω such that $|u|^p$ is integrable on Ω with respect to the Lebesgue measure. For $p = \infty$, we have $L_\infty(\Omega)$ denoting the set of all Lebesgue-measurable functions u defined on Ω such that $|u|$ has a finite essential supremum. The essential supremum of $|u|$ is defined as the infimum of the set of all positive real numbers M such that $|u| \leq M$ almost everywhere on Ω (see, for example, [13]).

The definition of the L_p norm, $p \geq 1$, for a function u defined in Ω is given by

$$||u||_{L_p(\Omega)} = \left(\int_{\mathbb{R}} |u(x)|^p \right)^{1/p}$$

and for $p = \infty$ is

$$||u||_{L_\infty(\Omega)} = \text{ess sup}_{x \in \Omega} |u(x)|.$$

Definition 1.1 Let $u(x) \in L_1(\Omega)$, $\Omega = (a, b)$, with $-\infty \leq a < b \leq \infty$. The left Riemann-Liouville fractional integral of order α in Ω, is given by

$$I_{a+}^\alpha u(x) = \frac{1}{\Gamma(\alpha)} \int_a^x u(t)(x - t)^{\alpha-1} dt.$$

The right Riemann-Liouville fractional integral of order α is given by

$$I_{b-}^\alpha u(x) = \frac{1}{\Gamma(\alpha)} \int_x^b u(t)(t - x)^{\alpha-1} dt.$$

Fractional integrals are defined for functions in $L_1(\Omega)$ existing almost everywhere. To prove this statement, we need to show that $I_{a+}^\alpha u \in L_1(\Omega)$ (see Exercise 1.2).

In particular we have the following result for the fractional integrals and finite domains.

Theorem 1.1 *If $u \in L_1(a, b)$ then*

$$||I_{a+}^{\alpha} u||_{L^1(a,b)} \leq \frac{(b-a)^{\alpha}}{\Gamma(\alpha+1)} ||u||_{L^1(a,b)},$$

$$||I_{b-}^{\alpha} u||_{L^1(a,b)} \leq \frac{(b-a)^{\alpha}}{\Gamma(\alpha+1)} ||u||_{L^1(a,b)}.$$

Proof The proof can be found in [119, pages 48–50]. □

Let

$$I_{a+}^{\alpha} u(a+) := \lim_{x \to a+} I_{a+}^{\alpha} u(x).$$

An interesting question to ask is for which functions $u(x)$ do we have

$$I_{a+}^{\alpha} u(a+) \neq 0?$$

If $u(x)$ is bounded, in particular at the lower limit, we have

$$I_{a+}^{\alpha} u(x) \leq \max_{x \in [a,b]} |u(x)| \frac{(x-a)^{\alpha}}{\Gamma(\alpha+1)}.$$

In this case $I_{a+}^{\alpha} u(a+) = 0$ and therefore this suggests we need to consider an unbounded function in order to get a non-zero result.

Let us consider the unbounded function at the lower limit $u(x) = (x-a)^{-\alpha}$. We have

$$I_{a+}^{\alpha} u(x) = \frac{1}{\Gamma(\alpha)} \int_a^x (t-a)^{-\alpha} (x-t)^{\alpha-1} dt.$$

From the result, related to the Gamma function properties,

$$\int_a^b (b-t)^{z-1} (t-a)^{w-1} dt = (b-a)^{z+w-1} \frac{\Gamma(z)\Gamma(w)}{\Gamma(z+w)} \tag{1.2}$$

we obtain

$$I_{a+}^{\alpha} u(x) = \frac{1}{\Gamma(\alpha)} (x-a)^0 \frac{\Gamma(\alpha)\Gamma(1-\alpha)}{\Gamma(1)} = 1.$$

Therefore,

$$I_{a+}^{\alpha} u(a+) = 1 \neq 0.$$

We proceed to present more information of interest regarding the fractional integrals. The first result is related to the fact that when $\alpha = 0$ the integral is the function under integration.

Theorem 1.2

(a) *If u is continuous then* $\lim_{\alpha \to 0} I_{a+}^{\alpha} u(x) = u(x)$;
(b) *If $u \in L_1(a,b)$, then* $\lim_{\alpha \to 0} I_{a+}^{\alpha} u(x) = u(x)$ *almost everywhere on* $[a,b]$.

Proof The proof of (a) can be found in [109, page 66] and the proof of (b) can be found in [119, pages 51–52]. □

The next result shows what happens with the composition of fractional integrals.

Theorem 1.3 *Let $\alpha > 0$ and $\beta > 0$. Then the equations*

$$(I_{a+}^{\alpha} I_{a+}^{\beta} u)(x) = I_{a+}^{\alpha+\beta} u(x),$$

$$(I_{b-}^{\alpha} I_{a+}^{\beta} u)(x) = I_{b-}^{\alpha+\beta} u(x)$$

are satisfied at almost every point $x \in [a,b]$ for $u(x) \in L_p(a,b)$, $1 \le p \le \infty$. If we have $\alpha + \beta > 1$ then these relations hold at almost every point of $[a,b]$.

Proof The proof is direct and is left as an exercise (see Exercise 1.6). It can also be found in [75, page 73] or [119, page 34]. □

1.2 Riemann-Liouville Fractional Derivatives

In order to have the existence of the fractional derivative guaranteed, it is not enough for u to be in L_1. Therefore, to define a set of functions for which these derivatives exist, we start to define the set of absolutely continuous functions.

The notion of absolute continuity allows the relationship between differentiation and integration. This relationship is commonly characterized by the fundamental theorem of calculus in the framework of Riemann integration, but with absolute continuity a similar relationship may be formulated in terms of Lebesgue integration.

Definition 1.2 A function $u(x)$ is called absolutely continuous on an interval $\overline{\Omega}$, if for any $\epsilon > 0$ there exists $\delta > 0$ such that for any finite set of pairwise nonintersecting intervals $[a_k, b_k] \subset \overline{\Omega}, k = 1, 2, \ldots, n$ such that $\sum_{k=1}^{n}(b_k - a_k) < \delta$ the inequality $\sum_{k=1}^{n} |u(b_k) - u(a_k)| < \delta$ holds. We denote the set of this functions by $AC(\overline{\Omega})$.

For the case when Ω is the infinite line, we will define $u \in AC(\mathbb{R})$ in the next chapter.

The definition by itself can be difficult to understand at first. But it means that absolutely continuous is stronger than continuous and weaker than Lipschitz continuous. It also means that a function $u \in AC(\overline{\Omega})$ can be written as

$$u(x) = u(a) + \int_a^x \phi(t)dt, \quad \text{for} \quad \phi \in L_1(\Omega) \tag{1.3}$$

and consequently that $u' \in L_1(\Omega)$ and it exists almost everywhere.

For the fractional differentiation, it is natural to introduce it as an operation inverse to integration as in the classical case. In the classical case suppose that for a given f we want to determine u such that

$$I_{a+}^1 u(x) = f(x).$$

Under the conditions that allow us to apply the fundamental theorem of calculus, we arrive at

$$u(x) = \frac{d}{dx} f(x).$$

It is a similar relation between integration and differentiation that will guide us to the definition of the Riemann-Liouville fractional derivative.

To illustrate that, let us assume that for a given f we want to determine u such that

$$I_{a+}^\alpha u(x) = f(x),$$

assuming the functions have all the conditions we need in the next calculations. This is called the Abel equation. Multiplying both sides by $(x - t)^{-\alpha}$ and integrating we have

$$\int_a^x \frac{I_{a+}^\alpha u(t)}{(x - t)^\alpha} dt = \int_a^x \frac{f(t)}{(x - t)^\alpha} dt.$$

Hence,

$$\int_a^x \frac{1}{(x - t)^\alpha} \int_a^t u(\tau)(t - \tau)^{\alpha-1} d\tau dt = \Gamma(\alpha) \int_a^x \frac{f(t)}{(x - t)^\alpha} dt.$$

Interchanging the order of integration in the left hand side we can write

$$\int_a^x u(\tau)d\tau \int_\tau^x \frac{dt}{(x - t)^\alpha (t - \tau)^{1-\alpha}} = \Gamma(\alpha) \int_a^x \frac{f(t)}{(x - t)^\alpha} dt.$$

Doing the change of variable $t = \tau + \eta(x - \tau)$ we have

$$\int_\tau^x (x - t)^{-\alpha}(t - \tau)^{-1+\alpha}dt = \int_0^1 \eta^{\alpha-1}(1 - \eta)^{-\alpha}d\eta.$$

Since it is known that

$$\int_0^1 \eta^{\alpha-1}(1 - \eta)^{-\alpha}d\eta = \Gamma(\alpha)\Gamma(1 - \alpha)$$

it follows

$$\int_a^x u(\tau)d\tau\, \Gamma(\alpha)\Gamma(1 - \alpha) = \Gamma(\alpha)\int_a^x \frac{f(t)}{(x - t)^\alpha}dt.$$

Therefore,

$$\int_a^x u(\tau)d\tau = \frac{1}{\Gamma(1 - \alpha)}\int_a^x \frac{f(t)}{(x - t)^\alpha}dt. \tag{1.4}$$

After differentiation we obtain

$$u(x) = \frac{1}{\Gamma(1 - \alpha)}\frac{d}{dx}\int_a^x \frac{f(t)}{(x - t)^\alpha}dt,$$

that can be rewritten as

$$u(x) = \frac{d}{dx}I_{a+}^{1-\alpha}f(x).$$

Let us state the rigorous result, known as the inversion result of Abel's equation and which proof can be found in [119, page 31]. Since we find it very interesting in the sense that shows the necessity of absolute continuity we present it here.

Theorem 1.4 *Let $\Omega = (a, b)$ and $\alpha \in (0, 1)$. The Abel integral equation*

$$I_{a+}^\alpha u(x) = f(x), \tag{1.5}$$

has a solution in $L_1(\Omega)$ if and only if $I_{a+}^{1-\alpha}f(x) \in AC(\overline{\Omega})$ and $I_{a+}^{1-\alpha}f(a) = 0$. Under these conditions, the solution is unique and given by

$$u(x) = \frac{d}{dx}I_{a+}^{1-\alpha}f(x). \tag{1.6}$$

Proof We start to prove the necessary condition. Let us assume that the Abel equation (1.5) has a solution for a function u belonging to $L_1(\Omega)$. The derivation above of the solution is valid, where the changing of order of the integration can be proved by the Fubini's Theorem and we can arrive at (1.4). Then, in view of (1.4), we obtain $I_{a+}^{1-\alpha} f(x) \in AC(\overline{\Omega})$ and $I_{a+}^{1-\alpha} f(a) = 0$.

For the sufficient condition, we have that $I_{a+}^{1-\alpha} f(x) \in AC(\overline{\Omega})$ and henceforth

$$\frac{d}{dx} I_{a+}^{1-\alpha} f(x) \in L_1(\Omega).$$

Then, the function given by (1.6) exists almost everywhere and belongs to $L_1(\Omega)$. We must show that is a solution of (1.5). With that purpose we substitute (1.6) in (1.5). Let us denote $g(x)$ the function

$$g(x) := I_{a+}^{\alpha} \frac{d}{dx} I_{a+}^{1-\alpha} f(x). \tag{1.7}$$

We want to show that $g(x) = f(x)$. We rewrite (1.7) as

$$\frac{1}{\Gamma(\alpha)} \int_a^x \frac{d}{dx} I_{a+}^{1-\alpha} f(t)(t-x)^{-\alpha} dt = g(x)$$

and is similar to (1.5). Therefore, by (1.6) we have

$$\frac{d}{dx} I_{a+}^{1-\alpha} f(x) = \frac{1}{\Gamma(1-\alpha)} \frac{d}{dx} \int_a^x \frac{g(t)}{(t-x)^{\alpha}} dt.$$

Hence,

$$\frac{d}{dx} I_{a+}^{1-\alpha} f(x) = \frac{d}{dx} I_{a+}^{1-\alpha} g(x).$$

Both functions are absolutely continuous, the function on the left hand side by hypothesis and the function on the right hand side because of (1.4). Therefore, the function $I_{a+}^{1-\alpha} f(x) - I_{a+}^{1-\alpha} g(x)$ is a constant. The condition of absolute continuity is essential in this conclusion. It can not be weakened by continuity since it is know that there are continuous functions, but not absolutely continuous functions, different from constant and having the derivative equal to zero almost everywhere, such as, the Cantor function [46]. If a function is absolutely continuous and its derivative is zero almost everywhere, then that function is constant. By hypothesis, we have $I_{a+}^{1-\alpha} f(a) = 0$ and by (1.7) we have $I_{a+}^{1-\alpha} g(a) = 0$, since this is an equation with a solution. Therefore the constant is zero and we have

$$I_{a+}^{1-\alpha} f(x) = I_{a+}^{1-\alpha} g(x).$$

Consequently,

$$\frac{1}{\Gamma(1-\alpha)} \frac{d}{dx} \int_a^x \frac{f(t) - g(t)}{(t-x)^{\alpha}} dt = 0.$$

This is an equation similar to (1.5) and therefore the uniqueness of its solution leads to $f(x) = g(x)$ and this completes the proof. $\qquad\square$

Consider the Abel equation, involving the right Riemann-Liouville fractional integral

$$I_{b-}^{\alpha} u(x) = f(x).$$

Hence,

$$\frac{1}{\Gamma(\alpha)} \int_{x}^{b} \frac{f(t)}{(t-x)^{1-\alpha}} dt = f(x), \quad x \le b.$$

In a similar way, we can obtain

$$u(x) = -\frac{1}{\Gamma(1-\alpha)} \frac{d}{dx} \int_{x}^{b} \frac{f(t)}{(t-x)^{\alpha}} dt.$$

From Theorem 1.4 the definitions of fractional derivatives, when $\alpha \in (0, 1)$, arrive naturally.

Definition 1.3 Let $\alpha \in (0, 1)$. The left Riemann-Liouville fractional derivative of order α is given by

$$D_{a+}^{\alpha} u(x) := \frac{d}{dx} I_{a}^{1-\alpha} u(x)$$

$$= \frac{1}{\Gamma(1-\alpha)} \frac{d}{dx} \int_{a}^{x} u(\xi)(x-\xi)^{-\alpha} d\xi.$$

Similarly, the right Riemann-Liouville fractional derivative of order α is given by

$$D_{b-}^{\alpha} u(x) := -\frac{d^{n}}{dx^{n}} I_{b}^{1-\alpha} u(x)$$

$$= \frac{-1}{\Gamma(1-\alpha)} \frac{d}{dx} \int_{x}^{b} u(\xi)(\xi-x)^{-\alpha} d\xi.$$

Before presenting the result that motivates the definition of fractional derivatives of higher order, we need to define the set AC^{n}, that is, the set of functions that are n times absolutely continuous.

Definition 1.4 We denote $AC^{n}(\overline{\Omega})$ the space of functions $u(x)$ which have continuous derivatives up to order $n-1$ on Ω with $u^{(n-1)}(x) \in AC(\overline{\Omega})$.

The next result generalizes the result presented in Theorem 1.4 regarding the Abel integral equation.

Theorem 1.5 *Let* $\Omega = (a, b)$ *and* $n - 1 < \alpha < n$. *The integral equation*

$$I_{a+}^{\alpha} u(x) = f(x), \tag{1.8}$$

has a solution in $L_1(\Omega)$ *if and only if* $I_{a+}^{n-\alpha} f(x) \in AC^n(\overline{\Omega})$ *and*

$$\frac{d^k}{dx^k}(I_{a+}^{n-\alpha} f)(a) = 0, \qquad k = 0, 1, 2, \ldots, n-1.$$

Under these conditions, the solution is unique and given by

$$u(x) = \frac{d^n}{dx^n} I_{a+}^{n-\alpha} f(x). \tag{1.9}$$

Proof The proof can be seen in [119, page 43]. $\qquad\qquad\square$

The previous result, motivates, the following definition of fractional derivative, for the general case.

Definition 1.5 Let $n \in N$ and $n-1 < \alpha < n$. The left Riemann-Liouville fractional derivative of order α is given by

$$D_{a+}^{\alpha} u(x) := \frac{d^n}{dx^n} I_a^{n-\alpha} u(x)$$

$$= \frac{1}{\Gamma(n-\alpha)} \frac{d^n}{dx^n} \int_a^x u(\xi)(x-\xi)^{n-1-\alpha} d\xi.$$

Similarly the right Riemann-Liouville fractional derivative of order α is given by

$$D_{b-}^{\alpha} u(x) := (-1)^n \frac{d^n}{dx^n} I_b^{n-\alpha} u(x)$$

$$= \frac{(-1)^n}{\Gamma(n-\alpha)} \frac{d^n}{dx^n} \int_x^b u(\xi)(\xi-x)^{n-1-\alpha} d\xi.$$

In the next result, we present under which conditions a fractional derivative exists.

Theorem 1.6 *Let* $u \in AC(\overline{\Omega})$ *and* $\alpha \in (0, 1)$. *Then* $D_{a+}^{\alpha} u(x)$ *exists almost everywhere.*

Proof The proof consists of showing the integral $I_{a+}^{1-\alpha} u(x)$ is absolutely continuous and therefore the derivative

$$D_{a+}^{\alpha} u(x) = \frac{d}{dx} I_{a+}^{1-\alpha} u(x)$$

is in $L_1(\Omega)$. We can write

$$u(x) = u(a) + \int_a^x u'(s)ds.$$

Therefore,

$$
\begin{aligned}
I_{a+}^{1-\alpha} u(x) &= \frac{1}{\Gamma(1-\alpha)} \int_a^x \left(u(a) + \int_a^t u'(s)ds \right) (x-t)^{-\alpha} dt \\
&= \frac{u(a)}{\Gamma(2-\alpha)}(x-a)^{1-\alpha} + \frac{1}{\Gamma(1-\alpha)} \int_a^x \int_a^t u'(s)(x-t)^{-\alpha} ds\, dt.
\end{aligned}
$$

$$(1.10)$$

The first term is absolutely continuous since

$$(x-a)^{1-\alpha} = (1-\alpha) \int_a^x (t-a)^{-\alpha} dt.$$

For the second integral, by changing the order of integration and a change of variables, we have

$$
\begin{aligned}
\frac{1}{\Gamma(1-\alpha)} \int_a^x \int_a^t u'(s)(x-t)^{-\alpha} ds\,dt &= \frac{1}{\Gamma(1-\alpha)} \int_a^x \int_s^x u'(s)(x-t)^{-\alpha} dt\,ds \\
&= \frac{1}{\Gamma(1-\alpha)} \int_a^x \int_s^x u'(s)(\eta-s)^{-\alpha} d\eta\,ds \\
&= \frac{1}{\Gamma(1-\alpha)} \int_a^x \int_a^t u'(s)(t-s)^{-\alpha} ds\,dt.
\end{aligned}
$$

$$(1.11)$$

Hence

$$\frac{1}{\Gamma(1-\alpha)} \int_a^x \int_a^t u'(s)(x-t)^{-\alpha} ds\,dt = \frac{1}{\Gamma(1-\alpha)} \int_a^x \left(\int_a^t u'(s)(t-s)^{-\alpha} ds \right) dt.$$

This is a primitive of an integrable function and therefore an absolutely continuous function.　　　　　　　　　　　　□

The next result is a sufficient condition for the existence of the Riemann-Liouville fractional derivative of higher order α, with $n-1 < \alpha < n$.

Theorem 1.7 *Let $u \in AC^n([a,b])$, $n = [\alpha]+1$, where $[\alpha]$ denotes the integer part of α. Then $D_{a+}^{\alpha} u$ exists almost everywhere.*

Proof The proof can be seen in [119, pages 39–40].　　　　　　　　　　□

At this moment, we can look at the definition of the fractional derivative of a function and be amazed by many things, including the non-locality and the kernel discontinuity. We can formulate the non-locality by a more astonishing fact: this derivative clearly depends on its full domain. This means that the value of a derivative, at a certain point, changes if the domain changes. Because of all of its particularities, it is understandable that the generalisation of the concept of integer order derivative has been the source of controversial discussions, centuries ago [117]. In the end, what it turns out is that this definition is more than a simple generalisation of the integer order derivative.

The next result is concerned with the question on how the derivatives of the same function, but different domains, relate to each other.

Theorem 1.8 *Let u be a bounded function and such that $u \in AC^n([a, c])$. Let $a < b < c$ and $n - 1 < \alpha < n$. For $b < x < c$, we have*

$$D^\alpha_{a+} u(x) = D^\alpha_{b+} u(x) + \psi(x),$$

where $\psi(x)$ is a function such that $\lim_{x \to \infty} \psi(x) = 0$.

Proof For $x > b$, we have that

$$\psi(x) = D^\alpha_{a+} u(x) - D^\alpha_{b+} u(x)$$

$$= \frac{1}{\Gamma(n - \alpha)} \frac{d^n}{dx^n} \int_a^b u(\tau)(x - \tau)^{n-1-\alpha} d\tau$$

$$= \frac{1}{\Gamma(n - \alpha)} \int_a^b u(\tau) \frac{d^n}{dx^n}[(x - \tau)^{n-1-\alpha}] d\tau$$

$$= \frac{1}{\Gamma(-\alpha)} \int_a^b u(\tau)(x - \tau)^{-1-\alpha} d\tau.$$

Since we are assuming that u is bounded, there exists an $M > 0$, such that $|u(x)| \le M$. Therefore,

$$|\psi(x)| \le \frac{M}{\Gamma(-\alpha)} \int_a^b (x - \tau)^{-1-\alpha} d\tau$$

$$= \frac{M}{\Gamma(1 - \alpha)}[(x - b)^{-\alpha} - (x - a)^{-\alpha}]$$

$$= \frac{M}{\Gamma(1 - \alpha)}(x - b)^{-\alpha}\left[1 - \left(\frac{x - b}{x - a}\right)^\alpha\right].$$

The limit of the right hand side goes to zero, when x goes to infinity. Hence $\lim_{x \to \infty} \psi(x) = 0$. $\qquad\qquad\square$

Another interesting result that can be illustrated by an example is that the fractional derivative of a bounded function can be unbounded.

The derivative of a constant is an unbounded solution, since

$$D_{0+}^{\alpha} 1 = \frac{1}{\Gamma(1-\alpha)} x^{-\alpha}, \quad \alpha > 0.$$

Other information of interest is to know how the fractional operators relate with the classical operators.

Theorem 1.9 *For $u \in AC^n([a, b])$, we have*

$$\lim_{\alpha \to n} D_{a+}^{\alpha} u(x) = u^{(n)}(x),$$

$$\lim_{\alpha \to n} D_{b-}^{\alpha} u(x) = (-1)^n u^{(n)}(x).$$

Proof When $\alpha \in (0, 1)$ and $\alpha \to 0$ we have $n = 1$ and

$$\lim_{\alpha \to 0} D_{a+}^{\alpha} u(x) = \lim_{\alpha \to 0} \frac{1}{\Gamma(1-\alpha)} \frac{d}{dx} \int_a^x u(\tau)(x-\tau)^{1-1-\alpha} d\tau.$$

Therefore,

$$\lim_{\alpha \to 0} D_{a+}^{\alpha} u(x) = \frac{1}{\Gamma(1)} \frac{d}{dx} \int_a^x u(\tau)(x-\tau)^0 d\tau$$

$$= u(x).$$

When $\alpha \to n - 1$ we have

$$\lim_{\alpha \to n-1} D_{a+}^{\alpha} u(x) = \lim_{\alpha \to n-1} \frac{1}{\Gamma(n-\alpha)} \frac{d^n}{dx^n} \int_a^x u(\tau)(x-\tau)^{n-1-\alpha} d\tau.$$

Therefore,

$$\lim_{\alpha \to n-1} D_{a+}^{\alpha} u(x) = \frac{1}{\Gamma(1)} \frac{d^n}{dx^n} \int_a^x u(\tau)(x-\tau)^0 d\tau$$

$$= u^{(n-1)}(x).$$

The result for the right Riemann-Liouville fractional derivative will follow in a similar way. $\qquad\square$

The properties related to the composition between integer order derivatives and fractional order derivatives are stated in the following result.

Theorem 1.10 *Let $n - 1 < \alpha < n$. If $u \in AC^{n+m}([a, b])$ then*

(a) $\quad \dfrac{d^m}{dx^m}\left[D_{a+}^{\alpha} u(x)\right] = D_{a+}^{\alpha+m} u(x),$

(b) $\quad \dfrac{d^m}{dx^m}\left[D_{b-}^{\alpha} u(x)\right] = (-1)^m D_{b-}^{\alpha+m} u(x),$

(c) $\quad D_{a+}^{\alpha}\left[\dfrac{d^m u}{dx^m}(x)\right] = D_{a+}^{\alpha+m} u(x) - \displaystyle\sum_{k=0}^{m-1} \dfrac{d^k u}{dx^k}(a) \dfrac{(x-a)^{k-\alpha-m}}{\Gamma(1+k-\alpha-m)},$

(d) $\quad D_{b-}^{\alpha}\left[\dfrac{d^m u}{dx^m}\right] = (-1)^m D_{b-}^{\alpha+m} u(x) - \displaystyle\sum_{k=0}^{m-1} (-1)^k \dfrac{d^k u}{dx^k}(a) \dfrac{(b-x)^{k-\alpha-m}}{\Gamma(1+k-\alpha-m)}.$

Proof

(a) Let $\beta \in (0, 1)$ and $k \geq 1$. We have $k - 1 < k - \beta < k$. Then

$$D_{a+}^{k-\beta} u(x) = \frac{1}{\Gamma(\beta)} \frac{d^k}{dx^k} \int_a^x u(\tau)(x-\tau)^{-1+\beta} d\tau.$$

Therefore,

$$D^m[D_{a+}^{k-\beta} u(x)] = \frac{1}{\Gamma(\beta)} \frac{d^{m+k}}{dx^{m+k}} \int_a^x u(\tau)(x-\tau)^{-1+\beta} d\tau$$

$$= D_{a+}^{m+k-\beta} u(x).$$

Let $\alpha := n - \beta$. We have

$$D^m[D_{a+}^{\alpha} u(x)] = D_{a+}^{m+\alpha} u(x).$$

(b) Similar to (a).

(c) This proof is more tricky than the previous proofs. First we observe that (see Exercise 1.10)

$$D_{a+}^{\alpha+m}[I^m u(x)] = D_{a+}^{\alpha} u(x). \tag{1.12}$$

Then, it follows that

$$D_{a+}^{\alpha+m}\left[I^m\left[\frac{d^m u}{dx^m}\right](x)\right] = D_{a+}^{\alpha}\left[\frac{d^m u}{dx^m}(x)\right]. \tag{1.13}$$

It is also easy to check that

$$I^m \left[\frac{d^m u}{dx^m} \right](x) = \frac{1}{(m-1)!} \int_a^x \frac{d^m u}{dx^m}(\tau)(x-\tau)^{m-1} d\tau$$

$$= u(x) - \sum_{k=0}^{m-1} \frac{d^k u}{dx^k}(a) \frac{(x-a)^k}{\Gamma(k+1)}. \tag{1.14}$$

If follows from (1.13) and (1.14) that

$$D_{a+}^\alpha \left[\frac{d^m u}{dx^m}(x) \right] = D_{a+}^{\alpha+m} \left[u(x) - \sum_{k=0}^{m-1} \frac{d^k u}{dx^k}(a) \frac{(x-a)^k}{\Gamma(k+1)} \right]$$

$$= D_{a+}^{\alpha+m} u(x) - \sum_{k=0}^{m-1} \frac{d^k u}{dx^k}(a) \frac{D_{a+}^{\alpha+m}\left[(x-a)^k\right]}{\Gamma(k+1)}. \tag{1.15}$$

We know that

$$D_{a+}^{\alpha+m}\left[(x-a)^k\right] = \frac{\Gamma(k+1)}{\Gamma(1+k-\alpha-m)}(x-a)^{k-\alpha-m}.$$

Inserting it in (1.15) we obtain the equality (c).

(d) In a similar way to (c) we can prove (d).

$\square$

The next result states a relation between fractional integration and fractional differentiation. It is a complementary result to Theorems 1.4 and 1.5.

Theorem 1.11 *Let $\alpha > 0$ and $n = [\alpha] + 1$. If $u \in L_1(a,b)$ and $I_{a+}^{n-\alpha} u \in AC^n([a,b])$, then the equalities*

$$(I_{a+}^\alpha D_{a+}^\alpha u)(x) = u(x) - \sum_{k=0}^{n-1} \frac{d^{n-k-1}}{dx^{n-k-1}} (I_{a+}^{n-\alpha} u)(a) \frac{(x-a)^{\alpha-k-1}}{\Gamma(\alpha-k)}$$

$$(I_{b-}^\alpha D_{b-}^\alpha u)(x) = u(x) - \sum_{k=0}^{n-1} (-1)^{n-k-1} \frac{d^{n-k-1}}{dx^{n-k-1}} (I_{b-}^{n-\alpha} u)(b) \frac{(b-x)^{\alpha-k-1}}{\Gamma(\alpha-k)}$$

hold almost everywhere on $[a,b]$.

Proof The proof concerning the left Riemann Liouville fractional integral and derivative can be seen in [119, page 45]. For the right Riemann Liouville fractional integral and derivative follows similarly. $\square$

The properties regarding the composition of fractional order derivatives are not generalizations of the integer order derivatives properties. A similar result is only valid for some particular cases as illustrated by the next result.

Theorem 1.12 *Let $\alpha, \beta > 0$ be such that $n - 1 < \alpha \leq n$, $m - 1 < \beta \leq m$ and $\alpha + \beta < n$. Let $u \in L_1(a, b)$, $I_{a+}^{m-\beta} u \in AC^m([a, b])$. Then*

$$D_{a+}^{\alpha} D_{a+}^{\beta} u(x) = D_{a+}^{\alpha+\beta} u(x) - \sum_{k=0}^{m-1} \frac{d^{m-k-1}}{dx^{m-k-1}} (I_{a+}^{m-\beta} u)(a) \frac{(x-a)^{-k-1-\alpha}}{\Gamma(-\alpha-k)}.$$

Proof We have

$$D_{a+}^{\alpha} D_{a+}^{\beta} u(x) = \left(\frac{d}{dx}\right)^n I_{a+}^{n-\alpha} D_{a+}^{\beta} u(x).$$

By Theorem 1.3 and since $\alpha + \beta < n$, it follows

$$D_{a+}^{\alpha} D_{a+}^{\beta} u(x) = \left(\frac{d}{dx}\right)^n I_{a+}^{n-\alpha-\beta} [I_{a+}^{\beta} D_{a+}^{\beta}] u(x)$$

$$= D_{a+}^{\alpha+\beta} [I_{a+}^{\beta} D_{a+}^{\beta}] u(x). \tag{1.16}$$

Since $u \in L_1(a, b)$ and $I_{a+}^{m-\beta} u \in AC^m([a, b])$, then by the previous theorem, we obtain

$$(I_{a+}^{\beta} D_{a+}^{\beta} u)(x) = u(x) - \sum_{k=0}^{m-1} \frac{d^{m-k-1}}{dx^{m-k-1}} (I_{a+}^{m-\beta} u)(a) \frac{(x-a)^{\beta-k-1}}{\Gamma(\beta-k)}.$$

Inserting into (1.16) then

$$D_{a+}^{\alpha} D_{a+}^{\beta} u(x) = D_{a+}^{\alpha+\beta} \left[u(x) - \sum_{k=0}^{m-1} \frac{d^{m-k-1}}{dx^{m-k-1}} (I_{a+}^{m-\beta} u)(a) \frac{(x-a)^{\beta-k-1}}{\Gamma(\beta-k)} \right].$$

Because

$$D_{a+}^{\alpha+\beta} [(x-a)^{\beta-k-1}] = \frac{\Gamma(\beta-k)}{\Gamma(-\alpha-k)} (x-a)^{-k-1-\alpha}$$

we obtain the stated result. $\square$

1.3 Caputo Fractional Derivatives

We introduce the definition of the Caputo fractional derivative and present several results that establish its relationship with the Riemann-Liouville fractional derivative. Some of these results will later be used in the next section to prove the equivalence between the Riemann-Liouville and the Grünwald-Letnikov fractional derivatives.

The Caputo fractional derivative is a more recent concept than the Riemann-Liouville fractional derivative definition.

Definition 1.6 The left Caputo fractional derivative is given by

$$c D_{a+}^{\alpha} u(x) = \frac{1}{\Gamma(n-\alpha)} \int_a^x u^{(n)}(\xi)(x-\xi)^{n-1-\alpha} d\xi.$$

Similarly the right Caputo fractional derivative is given by

$$c D_{b-}^{\alpha} u(x) = \frac{1}{\Gamma(n-\alpha)} \int_x^b u^{(n)}(\xi)(\xi-x)^{n-1-\alpha} d\xi.$$

This differential operator has the name Caputo since it is accepted that Caputo was the first using this operator in the context of applications [25] and the first that have studied some of its properties. Later it was incorporated by Caputo and Mainardi [26] in the framework of the theory of linear viscoelasticity.

This operator has different properties from the Riemann-Liouville fractional derivative. For instance, the Caputo fractional derivative is a linear operator that commutes, that is,

$$c D_{a+}^{\alpha} c D_{a+}^{\beta} u(x) = c D_{a+}^{\beta} c D_{a+}^{\alpha} u(x) = c D_{a+}^{\alpha+\beta} u(x), \quad \text{for all } \alpha, \beta > 0.$$

The Riemann-Liouville fractional derivative, as we have seen, although linear, in general does not commute.

Other interesting difference is that for the Riemann-Liouville fractional derivative of order $n-1 < \alpha < n$ we have that if

$$D_{0+}^{\alpha} f(x) = D_{0+}^{\alpha} g(x)$$

then

$$f(x) = g(x) + \sum_{j=1}^{n} c_j x^{\alpha-j}, \quad c_j \text{ constants,}$$

whereas for the Caputo fractional derivative if

$$c D_{0+}^{\alpha} f(x) = c D_{0+}^{\alpha} g(x),$$

then

$$f(x) = g(x) + \sum_{j=1}^{n} c_j x^{n-j}, \quad c_j \text{ constants.}$$

We give the Leibniz integration rule with variable limits of integration, in the following theorem, since it is difficult to find the rigorous formulation in literature. This rule will be needed in the proof of the next result, that relates the Caputo fractional derivative with the Riemann-Liouville fractional derivative.

Theorem 1.13 (Leibniz Integration Rule with Variable Limits of Integration)
Let Ω and X be open subsets of $\mathbb{R}$ and suppose that:

(a) For each $x \in X$, the function $u(x, \omega)$ is in $L_1(\Omega)$ and continuous in Ω.
(b) For almost all $\omega \in \Omega$, the partial derivative of u in order to x exists for all $x \in X$.
(c) There exists a function θ defined in Ω such that, for all $x \in X$ and almost every $\omega \in \Omega$, we have

$$\left| \frac{\partial u}{\partial x}(x, \omega) \right| \le \theta(\omega).$$

(d) The functions $\phi_1(x)$ and $\phi_2(x)$ are differentiable in X with values in Ω.

Then the function

$$g(x) = \int_{\phi_1(x)}^{\phi_2(x)} u(x, \omega) d\omega$$

is differentiable in X and

$$g'(x) = \int_{\phi_1(x)}^{\phi_2(x)} \frac{\partial u}{\partial x}(x, \omega) d\omega + u(x, \phi_2(x))\phi_2'(x) - u(x, \phi_1(x))\phi_1'(x).$$

In the next result, we relate the Riemann-Liouville fractional derivative with the Caputo fractional derivative.

Theorem 1.14 *Let $\alpha > 0$, $n = [\alpha] + 1$ and $u(x) \in AC^n([a, b])$. Then*

(a) $D_{a+}^{\alpha} u(x)$ exists almost everywhere and we have, for $x > a$,

$$_c D_{a+}^{\alpha} u(x) = D_{a+}^{\alpha} u(x) - \sum_{k=0}^{n-1} \frac{u^{(k)}(a)}{\Gamma(k - \alpha + 1)}(x - a)^{k-\alpha}.$$

(b) $D_{b-}^{\alpha} u(x)$ exists almost everywhere and we have, for $x < b$,

$$_c D_{b-}^{\alpha} u(x) = D_{b-}^{\alpha} u(x) - \sum_{k=0}^{n-1} \frac{u^{(k)}(b)}{\Gamma(k - \alpha + 1)}(b - x)^{k-\alpha}.$$

Proof

(a) We have

$$
D_{a+}^{\alpha} u(x) - \sum_{k=0}^{n-1} \frac{u^{(k)}(a)}{\Gamma(k-\alpha+1)}(x-a)^{k-\alpha}
$$

$$
= D_{a+}^{\alpha} \left[u(x) - \sum_{k=0}^{n-1} \frac{u^{(k)}(a)}{\Gamma(k+1)}(x-a)^{k} \right]
$$

$$
= \frac{1}{\Gamma(n-\alpha)} \left(\frac{d}{dx} \right)^{n} \int_{a}^{x} (x-\xi)^{n-\alpha-1}
$$

$$
\times \left(u(\xi) - \sum_{k=0}^{n-1} \frac{u^{(k)}(a)}{\Gamma(k+1)}(\xi-a)^{k} \right) d\xi.
$$

Doing integration by parts we obtain

$$
\int_{a}^{x} (x-\xi)^{n-\alpha-1} \left(u(\xi) - \sum_{k=0}^{n-1} \frac{u^{(k)}(a)}{\Gamma(k+1)}(\xi-a)^{k} \right) d\xi
$$

$$
= -\frac{(x-\xi)^{n-\alpha}}{n-\alpha} \left[u(\xi) - \sum_{k=0}^{n-1} \frac{u^{(k)}(a)}{\Gamma(k+1)}(\xi-a)^{k} \right] \Bigg|_{\xi=a}^{x}
$$

$$
+ \int_{a}^{x} \frac{(x-\xi)^{n-\alpha}}{n-\alpha} \frac{d}{d\xi} \left(u(\xi) - \sum_{k=0}^{n-1} \frac{u^{(k)}(a)}{\Gamma(k+1)}(\xi-a)^{k} \right) d\xi
$$

$$
= \int_{a}^{x} \frac{(x-\xi)^{n-\alpha}}{n-\alpha} \left(u'(\xi) - \sum_{k=1}^{n-1} \frac{u^{(k)}(a)}{\Gamma(k)}(\xi-a)^{k-1} \right) d\xi.
$$

By the Leibniz integration rule formulated previously

$$
\frac{d}{dx} \int_{a}^{x} \frac{(x-\xi)^{n-\alpha}}{n-\alpha} \left(u'(\xi) - \sum_{k=1}^{n-1} \frac{u^{(k)}(a)}{\Gamma(k)}(\xi-a)^{k-1} \right) d\xi
$$

$$
= \int_{a}^{x} (x-\xi)^{n-\alpha-1} \left[u'(\xi) - \sum_{k=1}^{n-1} \frac{u^{(k)}(a)}{\Gamma(k)}(\xi-a)^{k-1} \right] d\xi.
$$

Consequently,

$$\frac{1}{\Gamma(n-\alpha)} \left(\frac{d}{dx}\right)^n \int_a^x (x-\xi)^{n-\alpha-1} \left(u(\xi) - \sum_{k=0}^{n-1} \frac{u^{(k)}(a)}{\Gamma(k+1)}(\xi-a)^k\right) d\xi$$

$$= \frac{1}{\Gamma(n-\alpha)} \left(\frac{d}{dx}\right)^{n-1} \int_a^x (x-\xi)^{n-\alpha-1}$$

$$\times \left(u'(\xi) - \sum_{k=1}^{n-1} \frac{u^{(k)}(a)}{\Gamma(k)}(\xi-a)^{k-1}\right) d\xi.$$

Therefore, successively integration by parts alternated with the Leibniz integration rule leads to

$$D_{a+}^{\alpha}\left[u(x) - \sum_{k=0}^{n-1} \frac{u^{(k)}(a)}{\Gamma(k+1)}(x-a)^k\right]$$

$$= \frac{1}{\Gamma(n-\alpha)} \left(\frac{d}{dx}\right)^{n-1} \int_a^x (x-\xi)^{n-\alpha-1}$$

$$\times \left[u'(\xi) - \sum_{k=0}^{n-1} \frac{u^{(k)}(a)}{\Gamma(k+1)}(\xi-a)^{k-1}\right] d\xi$$

$$= \cdots = \frac{1}{\Gamma(n-\alpha)} \frac{d}{dx} \int_a^x (x-\xi)^{n-\alpha-1}\left[u^{(n-1)}(\xi) - u^{(n-1)}(a)\right] d\xi$$

$$= \frac{1}{\Gamma(n-\alpha)} \int_a^x (x-\xi)^{n-\alpha-1} u^{(n)}(\xi) d\xi$$

$$= {}_cD_{a+}^{\alpha} u(x).$$

(b) We can proceed similarly for the right Caputo fractional derivative.

$\square$

If α is not an integer, then the Caputo fractional derivative coincide with the Riemann-Liouville fractional derivative in the following cases.

If $u(a) = u'(a) = \cdots = u^{(n-1)}(a) = 0$, where $n = [\alpha] + 1$, then

$$_cD_{a+}^{\alpha} u(x) = D_{a+}^{\alpha} u(x)$$

and if $u(b) = u'(b) = \cdots = u^{(n-1)}(b) = 0$, where $n = [\alpha] + 1$, then

$$_cD_{b-}^{\alpha} u(x) = D_{b-}^{\alpha} u(x).$$

The next result follows very easily from Theorem 1.14 and clarifies the conditions under which the fractional Riemann-Liouville derivative vanishes at the initial point.

Theorem 1.15 *Let $\alpha > 0$, $n = [\alpha] + 1$ and $u \in AC^n([a, b])$.*

(a) The condition

$$\lim_{x \to a^+} D_{a+}^\alpha u(x) = 0$$

is equivalent to the conditions

$$u^{(k)}(a) = 0, \qquad k = 0, 1, 2, \ldots, n - 1.$$

(b) If the α-th derivative of u is equal to zero when x goes to the terminal value a, then all derivatives of order β, for $0 < \beta < \alpha$, are also equal to zero when x goes to the terminal value a, that is,

$$\lim_{x \to a^+} D_{a+}^\beta u(x) = 0.$$

Proof

(a) If we have $u^{(k)}(a) = 0$, $k = 0, 1, 2, \ldots, n - 1$, then by applying the limit, x goes to a, on both sides of the expression of Theorem 1.14 we obtain $\lim_{x \to a^+} D_{a+}^\alpha u(x) = 0$. Conversely, if we have $\lim_{x \to a^+} D_{a+}^\alpha u(x) = 0$ then multiplying both sides of the expression in Theorem 1.14 by $(x - a)^{\alpha - j}$, $j = n - 1, n - 2, \ldots, 2, 1, 0$ and taking the limit, as x goes to a, we obtain $u^{(k)}(a) = 0$, $k = 0, 1, 2, \ldots, n - 1$.

(b) This immediately follows from (a).

$\square$

1.4 Grünwald-Letnikov Fractional Derivatives

We introduce the definition of the Grünwald-Letnikov fractional derivative and discuss how this derivative relates to the Riemann-Liouville fractional derivative. Most results presented here have also been discussed in [43, 75, 109, 119].

Definition 1.7 The Grünwald-Letnikov fractional derivative is given by

$$_{GL}D_{a+}^\alpha u(x) = \lim_{\Delta x \to 0} \frac{1}{\Delta x^\alpha} \sum_{k=0}^{\left[\frac{x-a}{\Delta x}\right]} (-1)^k \binom{\alpha}{k} u(x - k\Delta x), \qquad (1.17)$$

$$_{GL}D_{b-}^\alpha u(x) = \lim_{\Delta x \to 0} \frac{1}{\Delta x^\alpha} \sum_{k=0}^{\left[\frac{b-x}{\Delta x}\right]} (-1)^k \binom{\alpha}{k} u(x + k\Delta x), \qquad (1.18)$$

where

$$(-1)^k \binom{\alpha}{k} = (-1)^k \frac{\Gamma(\alpha + 1)}{\Gamma(k + 1)\Gamma(\alpha - k + 1)}$$

$$= \frac{\Gamma(k - \alpha)}{\Gamma(-\alpha)\Gamma(k + 1)}.$$

The Grünwald-Letnikov fractional derivative can be seen as a generalization of the ordinary discretization formulas for integer order derivatives. If we consider the domain $\mathbb{R}$ the sum in (1.17) is a series. This series converges absolutely and uniformly for each $\alpha > 0$ and for every bounded function $u(x)$.

Before stating the result that relates the Riemann-Liouville fractional derivative to the Grünwald-Letnikov fractional derivative, we introduce the Grünwald-Letnikov definition of fractional integration and a theorem related to it.

Definition 1.8 The Grünwald-Letnikov fractional integral is given by

$$_{GL}I_{a+}^{\alpha}u(x) = \lim_{\Delta x \to 0} \Delta x^{\alpha} \sum_{k=0}^{\left[\frac{x-a}{\Delta x}\right]} (-1)^k \binom{-\alpha}{k} u(x - k\Delta x), \qquad (1.19)$$

$$_{GL}I_{b-}^{\alpha}u(x) = \lim_{\Delta x \to 0} \Delta x^{\alpha} \sum_{k=0}^{\left[\frac{b-x}{\Delta x}\right]} (-1)^k \binom{-\alpha}{k} u(x + k\Delta x), \qquad (1.20)$$

where

$$(-1)^k \binom{-\alpha}{k} = \frac{\Gamma(k + \alpha)}{\Gamma(\alpha)\Gamma(k + 1)}.$$

The next result shows that the Grünwald-Letnikov fractional integral $_{GL}I_{a+}^{\alpha}u$ is just the same as the Riemann-Liouville fractional integral.

Theorem 1.16 *Let $\alpha > 0$ and let $u \in L_1(a, b)$. Then $_{GL}I_{a+}^{\alpha}u(x)$ exists for almost all x and*

$$_{GL}I_{a+}^{\alpha}u(x) = \frac{1}{\Gamma(\alpha)} \int_a^x u(\xi)(x - \xi)^{\alpha-1}d\xi.$$

Proof The proof is left as an exercise (see Exercise 1.11).   □

We proceed now to the main result that highlights the relation between the three definitions of fractional derivative, that we have been discussing in this chapter.

Theorem 1.17 *Let $n - 1 < \alpha < n$ and $u \in AC^n([a, b])$. Then, for almost every $x \in (a, b]$,*

$$_C D_{a+}^{\alpha} u(x) + \sum_{k=0}^{n-1} \frac{u^{(k)}(a)}{\Gamma(k - \alpha + 1)} (x - a)^{k-\alpha} = {_{GL}} D_{a+}^{\alpha} u(x).$$

Proof The proof consists of three steps:

Step 1: Define

$$\Delta_{\Delta x} u(x) := u(x) - u(x - \Delta x),$$
$$\Delta_{\Delta x}^k u(x) := \Delta_{\Delta x}^{k-1} u(x) - \Delta_{\Delta x}^{k-1} u(x - \Delta x).$$

For $x - a = M \Delta x$,

$$\frac{1}{\Delta x^{\alpha}} \sum_{k=0}^{M} (-1)^k \binom{\alpha}{k} u(x - k \Delta x)$$

$$= \frac{1}{\Delta x^{\alpha}} \sum_{r=0}^{n-1} (-1)^{M-r} \binom{\alpha - r - 1}{M - r} \Delta_{\Delta x}^r u(a + r \Delta x)$$

$$+ \frac{1}{\Delta x^{\alpha}} \sum_{r=0}^{M-n} (-1)^r \binom{\alpha - n}{r} \Delta_{\Delta x}^n u(a - r \Delta x). \tag{1.21}$$

Step 2:

$$\lim_{\Delta x \to 0} \frac{1}{\Delta x^{\alpha}} \sum_{r=0}^{n-1} (-1)^{M-r} \binom{\alpha - r - 1}{M - r} \Delta_{\Delta x}^r u(a + r \Delta x)$$

$$= \sum_{k=0}^{n-1} \frac{u^{(k)}(a)}{\Gamma(k - \alpha + 1)} (x - a)^{k-\alpha}.$$

Step 3:

$$\lim_{\Delta x \to 0} \frac{1}{\Delta x^{\alpha}} \sum_{r=0}^{M-n} (-1)^r \binom{\alpha - n}{r} \Delta_{\Delta x}^n u(x - r \Delta x)$$

$$= \frac{1}{\Gamma(n - \alpha)} \int_a^x (x - t)^{n-\alpha-1} u^{(n)}(t) dt.$$

Proof of Step 1: Considering that

$$\binom{\alpha}{k} = \binom{\alpha-1}{k} + \binom{\alpha-1}{k-1} \tag{1.22}$$

we can write

$$\frac{1}{\Delta x^\alpha} \sum_{k=0}^{M} (-1)^k \binom{\alpha}{k} u(x - k\Delta x) = \frac{1}{\Delta x^\alpha} \sum_{k=0}^{M} (-1)^k \binom{\alpha-1}{k} u(x - k\Delta x)$$

$$+ \frac{1}{\Delta x^\alpha} \sum_{k=0}^{M} (-1)^k \binom{\alpha-1}{k-1} u(x - k\Delta x).$$

For the second sum we have

$$\sum_{k=0}^{M} (-1)^k \binom{\alpha-1}{k-1} u(x - k\Delta x) = - \sum_{k=0}^{M-1} (-1)^k \binom{\alpha-1}{k} u(x - (k+1)\Delta x).$$

Therefore,

$$\frac{1}{\Delta x^\alpha} \sum_{k=0}^{M} (-1)^k \binom{\alpha}{k} u(x - k\Delta x)$$

$$= \frac{1}{\Delta x^\alpha} (-1)^M \binom{\alpha-1}{M} u(a)$$

$$+ \frac{1}{\Delta x^\alpha} \sum_{k=0}^{M-1} (-1)^k \binom{\alpha-1}{k} (u(x - k\Delta x) - u(x - (k+1)\Delta x)).$$

We obtain

$$\frac{1}{\Delta x^\alpha} \sum_{k=0}^{M} (-1)^k \binom{\alpha}{k} u(x - k\Delta x)$$

$$= \frac{1}{\Delta x^\alpha} (-1)^M \binom{\alpha-1}{M} u(a)$$

$$+ \frac{1}{\Delta x^\alpha} \sum_{k=0}^{M-1} (-1)^k \binom{\alpha-1}{k} \Delta_{\Delta x} u(x - k\Delta x).$$

Repeating this procedure $n-1$ times, using (1.22), we obtain (1.21).

Proof of Step 2: To determine the limit

$$\lim_{\Delta x \to 0} \frac{1}{\Delta x^\alpha} \sum_{r=0}^{n-1} (-1)^{M-r} \binom{\alpha - r - 1}{M - r} \Delta_{\Delta x}^r u(a + r\Delta x)$$

we rewrite the r-term of the sum as

$$\frac{1}{\Delta x^\alpha} (-1)^{M-r} \binom{\alpha - r - 1}{M - r} \Delta_{\Delta x}^r u(a + r\Delta x)$$

$$= (-1)^{M-r} \binom{\alpha - r - 1}{M - r} (M - r)^{\alpha - r} \left(\frac{M}{M - r}\right)^{\alpha - r} (M\Delta x)^{r-\alpha} \frac{\Delta_{\Delta x}^r u(a + r\Delta x)}{\Delta x^r}.$$

Since $M\Delta x = x - a$ and

$$\lim_{M \to \infty} (-1)^{M-r} \binom{\alpha - r - 1}{M - r} (M - r)^{\alpha - r} = \frac{1}{\Gamma(-\alpha + r + 1)},$$

$$\lim_{M \to \infty} \left(\frac{M}{M - r}\right)^{\alpha - r} = 1,$$

$$\lim_{\Delta x \to \infty} \frac{\Delta_{\Delta x}^r u(a + r\Delta x)}{\Delta x^r} = u^{(r)}(a)$$

it follows the equality (1.22).

Proof of Step 3: We start to notice that

$$\lim_{\Delta x \to 0} \frac{1}{\Delta x^\alpha} \sum_{r=0}^{M-n} (-1)^r \binom{\alpha - n}{r} \Delta_{\Delta x}^n u(x - r\Delta x)$$

$$= \lim_{\Delta x \to 0} \frac{(\Delta x)^n}{(\Delta x)^\alpha} \sum_{r=0}^{\left[\frac{x-a}{\Delta x}\right]} (-1)^r \binom{\alpha - n}{r} \frac{\Delta_{\Delta x}^n u(x - r\Delta x)}{(\Delta x)^n}$$

$$= \lim_{\Delta x \to 0} \frac{(\Delta x)^n}{(\Delta x)^\alpha} \sum_{r=0}^{\left[\frac{x-a}{\Delta x}\right]} (-1)^r \binom{\alpha - n}{r} \lim_{\Delta x_1 \to 0} \frac{\Delta_{\Delta x_1}^n u(x - r\Delta x)}{(\Delta x_1)^n}$$

$$= \lim_{\Delta x \to 0} (\Delta x)^{n-\alpha} \sum_{r=0}^{\left[\frac{x-a}{\Delta x}\right]} (-1)^r \binom{\alpha - n}{r} u^{(n)} u(x - r\Delta x).$$

The last two equalities are due to the fact that all the previous limits exist and that

$$\lim_{\Delta x_1 \to 0} \frac{\Delta_{\Delta x_1}^n u(x)}{(\Delta x_1)^n} = u^{(n)}(x).$$

Now using the previous result in Theorem 1.16 we have

$$\lim_{\Delta x \to 0} (\Delta x)^{n-\alpha} \sum_{r=0}^{\left[\frac{x-a}{\Delta x}\right]} (-1)^r \binom{\alpha - n}{r} u^{(n)} u(x - r\Delta x)$$

$$= \frac{1}{\Gamma(n-\alpha)} \int_a^x u^{(n)}(t)(x-t)^{n-\alpha-1} dt.$$

$\square$

From the previous theorem and Theorem 1.14 in Sect. 1.3, we obtain the following relation between the Riemann-Liouville and the Grünwald-Letnikov fractional derivatives.

Corollary 1.1 *Let $n - 1 < \alpha < n$ and $u \in AC^n([a, b])$. Then, for almost every $x \in [a, b]$,*

$$_{GL}D_{a+}^{\alpha} u(x) = D_{a+}^{\alpha} u(x).$$

We can also prove similar statements as the previous theorem and corollary for functions given on the axis or half-axis provided that the functions behave appropriately at infinity.

1.5 Exercises and Solutions

Exercises

1.1 Prove the Cauchy formula

$$I_{a+}^n u(x) = \frac{1}{(n-1)!} \int_a^x u(t)(x-t)^{n-1} dt.$$

1.2 Prove that the fractional integral $I_{a+}^{\alpha} u$ exists almost everywhere, when $u \in L_1(\Omega)$, for $\Omega = (a, b)$.

1.3 Give an example of a function for which the fractional integral $I_{a+}^{1-\alpha} u(a) \neq 0$, when $\alpha \in (0, 1)$.

1.4 Prove that if $u \in L_\infty(a, b)$ then

$$||I_{a+}^{\alpha} u||_{L_\infty(a,b)} \leq \frac{(b-a)^{\alpha}}{\Gamma(\alpha + 1)} ||u||_{L_\infty(a,b)},$$

$$||I_{b-}^{\alpha} u||_{L_\infty(a,b)} \leq \frac{(b-a)^{\alpha}}{\Gamma(\alpha + 1)} ||u||_{L_\infty(a,b)}.$$

1.5 Show that, for $u(x) = x$,

$$\lim_{\alpha \to 0} I_{0+}^{\alpha} u(x) = u(x).$$

1.6 Prove Theorem 1.3.

1.7 (a) For $\alpha > 0$, compute $D_{0+}^{\alpha} u(x)$, when $u(x) = 1$.

   (b) For $\alpha > 0$, compute $D_{0+}^{\alpha} u(x)$, when $u(x) = x^2$.

   (c) For $\alpha > 0$, compute $D_{0+}^{\alpha} u(x)$, when $u(x) = x^{\gamma}$, for $\gamma > 0$.

1.8 (a) Determine the solution of the boundary value problem $u''(x) = 0$ defined in $(0, 1)$ with the boundary conditions $u(0) = 0$ and $u(1) = 1$.

   (b) Determine the solution of the boundary value problem $D_{0+}^{\alpha} u(x) = 0$, for $\alpha \in (1, 2)$ defined in $(0, 1)$ with the boundary conditions $u(0) = 0$ and $u(1) = 1$. What can you say about this problem if the boundary at $x = 0$ is not zero?

1.9 Prove that for $n - 1 < \alpha < n$

$$D_{a+}^{\alpha} u(x) = 0 \text{ if } u(x) = (x - a)^{\alpha - k}, \ k = 0, 1, \dots, n.$$

1.10 Let $n - 1 < \alpha < n$. Prove that if $u \in AC^{n+m}([a, b])$, then $D_{a+}^{\alpha+m}[I^{m} u(x)] = D_{a+}^{\alpha} u(x)$.

1.11 Prove Theorem 1.16.

Solutions

1.1 Prove the equality by induction, that is, it is true for $n = 1$ and for $n + 1$ assuming it is true for n.

1.2 Prove that $I_{a+}^{n} u$ belongs to $L_1(\Omega)$, using the generalised Minkowsky inequality.

   The generalised Minkowsky inequality [119, page 9]: Let $\Omega_1 = (a, b)$ and $\Omega_2 = (c, d)$, $-\infty \le a < b \le \infty$, $-\infty \le c < d \le \infty$, and let $f(x, y)$ be a measurable function defined on $\Omega_1 \times \Omega_2$. If at least one of the integrals

$$\int_{\Omega_1} dx \int_{\Omega_2} f(x, y) dy, \quad \int_{\Omega_2} dy \int_{\Omega_1} f(x, y) dx, \quad \int_{\Omega_1 \times \Omega_2} f(x, y) dx dy$$

is absolutely convergent then they coincide and we have

$$\left\{ \int_{\Omega_1} dx \left| \int_{\Omega_2} f(x, y) dy \right|^{p} \right\}^{1/p} \le \int_{\Omega_2} dy \left\{ \int_{\Omega_1} |f(x, y)|^{p} dx \right\}^{1/p}.$$

1.3 Choose a function that is not in $L_1(\Omega)$.

1.4 Start to prove that, for each $x \in (a, b)$,

$$\left| I_{a+}^{n} u(x) \right| \leq \|u\|_{L_\infty(a,b)} \frac{1}{\Gamma(\alpha)} \frac{(x-a)^\alpha}{\alpha}.$$

1.5 Use property (1.2).
1.6 Write the expression of $I_{a+}^{\alpha} I_{a+}^{\beta} u$. Change the order of integration by Fubini's theorem and set a proper change of variables to obtain the final equality.
1.7 In general we have, for $\gamma > -1$,

$$D_{0+}^{\alpha}(x^\gamma) = \frac{\Gamma(\gamma+1)}{\Gamma(\gamma+1-\alpha)} x^{\gamma-\alpha}.$$

1.8 (a) We want to determine a function u, defined on $(0, 1)$, and such that

$$u''(x) = 0.$$

This has a very simple solution

$$u(x) = Ax + B.$$

For $u(0) = 0$ and $u(1) = 1$ we obtain $B = 0$ and $A = 1$. Therefore $u(x) = x$.
   (b) Now, let us consider the problem of determining $u(x)$ on $(0, 1)$ such that

$$D_{0+}^{\alpha} u(x) = 0, \quad \text{for} \quad \alpha \in (1, 2).$$

We can write it as

$$\frac{d^2}{dx^2} (I_{0+}^{2-\alpha} u)(x) = 0.$$

Then

$$I_{0+}^{2-\alpha} u(x) = Ax + B =: \psi(x)$$

By Theorem 1.4, this equation has a solution in $L_1(\Omega)$ if and only if

$$I_{0+}^{\alpha-1} \psi(x) \in AC(\overline{\Omega}) \quad \text{and} \quad I_{0+}^{\alpha-1} \psi(0) = 0.$$

We have that

$$I_{0+}^{\alpha-1} \psi(x) = \frac{1}{\Gamma(\alpha-1)} \int_a^x \psi(\tau)(x-\tau)^{\alpha-2} d\tau = \frac{Ax^\alpha}{\Gamma(\alpha+1)} + \frac{Bx^{\alpha-1}}{\Gamma(\alpha)}.$$

We can confirm that $I_{0+}^{\alpha-1}\psi(x) \in AC(\overline{\Omega})$ and $I_{0+}^{\alpha-1}\psi(0) = 0$. Under these conditions, according to Theorem 1.4, we have

$$u(x) = \frac{d}{dx}\left(I_{0+}^{\alpha-1}\psi(x)\right) = \frac{Ax^{\alpha-1}}{\Gamma(\alpha)} + \frac{Bx^{\alpha-2}}{\Gamma(\alpha-1)}.$$

However, this is a solution of the integral equation, and not necessarily of the problem we are trying to solve. If we want a solution such that $u(0) = 0$ we need to have $B = 0$ in the previous expression which is equivalent to say that

$$I_{0+}^{2-\alpha}u(0) = \psi(0).$$

In this case, for $u(1) = 1$, we obtain the final solution

$$u(x) = x^{\alpha-1}.$$

A solution that is not zero at $x = 0$ will be an unbounded solution and not defined at $x = 0$.

1.9   Compute it similarly to Exercise 1.7.

1.10 First observe that by definition the Riemann-Liouville fractional derivative for $n - 1 < \alpha < n$ satisfy

$$D_{a+}^{\alpha+m}(I^m u(x)) = \frac{d^{n+m}}{dx^{n+m}}(I^{n+m-(\alpha+m)} I^m u(x)).$$

Then use the property of fractional integrals (see Theorem 1.3) to write

$$\frac{d^{n+m}}{dx^{n+m}}(I^{n-\alpha} I^m u(x)) = \frac{d^n}{dx^n}(I^{n-\alpha} u(x)) = D_{a+}^{\alpha} u(x).$$

1.11 Observe that the function $\xi^{\alpha-1}u(x - \xi)$ is integrable for almost every x and that in this case the integral can be written as the limit of a sum. The detailed proof can be found in [119, pages 387–388].

Chapter 2
A Governing Equation for Lévy Flights

Lévy flights and the associated distributions characterize anomalous diffusive processes. The access to the analytical expressions of these distributions is often difficult. However, to study these distributions a fractional differential equation can be considered, since they are the fundamental solutions of these equations. We start by giving a brief introduction to the Riemann-Liouville fractional derivatives and Fourier transform of a function defined in the real line. Subsequently properties of the Fourier transform of the Riemann-Liouville fractional derivatives are presented. They are necessary to show how from the Lévy alpha-stable distributions we arrive at a fractional differential equation. We end with the presentation of analytical solutions for some fractional diffusion equations and discuss its properties.

2.1 Fractional Integrals and Derivatives on the Real Axis

In this section, we revisit the definitions of fractional integrals and derivatives for functions defined on the real line in order to align with the definition of the Fourier transform and to account for the subtle differences that arise in this setting when compared to the bounded interval case.

We present the definition of fractional integral and fractional derivative when we have a function defined in the real line.

Definition 2.1 Let $u \in L_1(\mathbb{R})$. The left Riemann-Liouville fractional integral of order α is given by

$$I_+^\alpha u(x) = \frac{1}{\Gamma(\alpha)} \int_{-\infty}^{x} u(t)(x-t)^{\alpha-1} dt.$$

E. Sousa, *Finite Difference Methods for Fractional Diffusion Equations*, Lecture Notes in Mathematics 2389, https://doi.org/10.1007/978-3-032-11222-4_2

The right Riemann-Liouville fractional integral of order α is given by

$$I_-^\alpha u(x) = \frac{1}{\Gamma(\alpha)} \int_x^\infty u(t)(t-x)^{\alpha-1}dt.$$

Fractional integrals are defined for functions in $L_1(\mathbb{R})$ existing almost everywhere (see Exercise 2.1).

Definition 2.2 Let $n \in N$ and $n-1 < \alpha < n$. The left Riemann-Liouville fractional derivative of order α is given by

$$D_+^\alpha u(x) := \frac{d^n}{dx^n} I_+^{n-\alpha} u(x)$$

$$= \frac{1}{\Gamma(n-\alpha)} \frac{d^n}{dx^n} \int_{-\infty}^x u(\xi)(x-\xi)^{n-1-\alpha}d\xi.$$

The right Riemann-Liouville fractional derivative of order α is given by

$$D_-^\alpha u(x) := (-1)^n \frac{d^n}{dx^n} I_-^{n-\alpha} u(x)$$

$$= \frac{(-1)^n}{\Gamma(n-\alpha)} \frac{d^n}{dx^n} \int_x^\infty u(\xi)(\xi-x)^{n-1-\alpha}d\xi.$$

Similarly to what we have done for the case of functions defined in a finite interval, the spaces where these derivatives are defined can be characterized in terms of the absolute continuity of the fractional integrals. In Chap. 1, we have defined absolute continuity in an interval. A function $u(x)$ is called absolutely continuous on an interval $\overline{\Omega}$, if for any $\epsilon > 0$ there exists $\delta > 0$ such that for any finite set of pairwise nonintersecting intervals $[a_k, b_k] \subset \overline{\Omega}, k = 1, 2, \ldots, n$ such that $\sum_{k=1}^n (b_k - a_k) < \delta$, the inequality $\sum_{k=1}^n |u(b_k) - u(a_k)| < \delta$ holds.

We now define absolute continuity in the real line, according to [119, page 130].

Definition 2.3 We say $u \in AC(\mathbb{R})$, if u is absolutely continuous on any finite interval and has bounded variation on the closed real line $\mathbb{R}$.

The previous statement that the function u belongs to $AC(\mathbb{R})$ is equivalent to its representability in the form (see [119, page 130]),

$$u(x) = c + \int_{-\infty}^x \phi(t)dt, \quad \text{for} \quad \phi \in L_1(\mathbb{R}). \tag{2.1}$$

Note that when $\Omega = \mathbb{R}$, it does not necessarily mean that $u' \in L_1(\mathbb{R})$.

Next, we define the absolutely continuous functions of order n.

Definition 2.4 We denote $AC^n(\mathbb{R})$ the space of functions $u(x)$ which have continuous derivatives up to order $n - 1$ on $\mathbb{R}$ with $u^{(n-1)}(x) \in AC(\mathbb{R})$.

Many of the properties given in the previous chapter, for fractional integrals and derivatives, can be adjusted to the case of having the fractional integrals and derivatives defined in the real line as illustrated by the next result.

Theorem 2.1 *Let $\alpha \in (0, 1)$. The integral equation*

$$I_+^\alpha u(x) = f(x), \tag{2.2}$$

has a solution in $L_1(\mathbb{R})$ if and only if $I_+^{1-\alpha} f(x) \in AC(\mathbb{R})$ and $I_+^{1-\alpha} f(-\infty) = 0$. Under these conditions, the solution is unique and given by

$$u(x) = \frac{d}{dx} I_+^{1-\alpha} f(x). \tag{2.3}$$

Proof The proof is similar to the case of a finite interval (see Theorem 1.4 in Chap. 1). □

A similar result follows for the case of the right fractional integral.

2.2 The Fourier Transform

In this section, we introduce some properties of the Fourier transform for functions in L_1. We need the concept of Fourier transform to explain how do we go from the idea of Lévy flights to the fractional diffusion equation.

Definition 2.5 Let $u \in L_1(\mathbb{R})$. The Fourier transform of u is defined by

$$\mathscr{F}(u)(\xi) = \frac{1}{\sqrt{2\pi}} \int_\mathbb{R} e^{ix\xi} u(x)dx.$$

The integral involved in the definition of the Fourier transform converges absolutely for functions $u \in L_1(\mathbb{R})$, that is, $\mathscr{F}(u)$ of $u \in L_1$ exists for all u. The Fourier transform of the function $u \in L_1$ is a bounded continuous function satisfying

$$|\mathscr{F}(u)| \leq ||u||_1. \tag{2.4}$$

The Fourier transform of u tends to zero as $|x| \to \infty$ by the Riemann-Lebesgue theorem. The rate of decrease at infinity of $\mathscr{F}(u)$ is connected with the smoothness of the function u.

In order to present the convolution property of the Fourier transform, we define the convolution product as

$$(f * g)(x) = \frac{1}{\sqrt{2\pi}} \int_{\mathbb{R}} f(x - \tau)g(\tau)d\tau.$$

We have the following result.

Proposition 2.1 *If* $f, g \in L_1(\mathbb{R})$, *then*

$$\mathscr{F}(f * g)(\xi) = \mathscr{F}(f)(\xi)\mathscr{F}(g)(\xi).$$

Proof The convolution $f * g$ exists almost everywhere as a function of L_1 (see Exercise 2.2). Therefore we have, for every $\xi \in \mathbb{R}$,

$$\mathscr{F}(f * g)(\xi) = \frac{1}{\sqrt{2\pi}} \int_{\mathbb{R}} \left\{ \frac{1}{\sqrt{2\pi}} \int_{\mathbb{R}} f(x - \tau)g(\tau)d\tau \right\} e^{i\xi x} dx$$

$$= \frac{1}{\sqrt{2\pi}} \int_{\mathbb{R}} g(\tau)e^{i\xi\tau}d\tau \left\{ \frac{1}{\sqrt{2\pi}} \int_{\mathbb{R}} f(x - \tau)e^{-i\tau(x-\tau)} \right\} dx$$

$$= \mathscr{F}(f)(\xi)\mathscr{F}(g)(\xi).$$

$\square$

In the next result, we consider the inversion operator, that is, how to reconstruct the original function u from the value of $\mathscr{F}(u)$.

Theorem 2.2 *If* u *and* $\mathscr{F}(u) \in L_1(\mathbb{R})$ *then*

$$u(x) = \frac{1}{\sqrt{2\pi}} \int_{\mathbb{R}} e^{-ix\xi} \mathscr{F}(u)(\xi)d\xi,$$

for almost all $x \in \mathbb{R}$. *Therefore,* u *is equal almost everywhere to a function in* $L_1(\mathbb{R}) \cap C(\mathbb{R})$. *If* u *is continuous on* $\mathbb{R}$, *then the inversion formula holds everywhere.*

Proof The proof requires the introduction of several complex concepts and results that would only be used to this purpose. Therefore, we do not present it here. It can be seen in Section 5.1.2 of [23, pages 190–194]. $\square$

We need to pay attention to this inversion formula, since the Fourier transform $\mathscr{F}(u)$ of $u \in L_1$ does not necessarily belong to L_1.

We can now define the inverse Fourier transform.

Definition 2.6 Let $v \in L_1(\mathbb{R})$. The Fourier transform of v is defined by

$$\mathscr{F}^{-1}(v)(x) = \frac{1}{\sqrt{2\pi}} \int_{\mathbb{R}} e^{-ix\xi} v(\xi)d\xi.$$

We proceed to the presentation of the results of the Fourier transforms of the fractional derivatives. We start to state the results for the derivatives of integer order and then continue to the fractional order derivatives.

Proposition 2.2 *Let $u \in L_1(\mathbb{R}) \cap AC^{(n-1)}(\mathbb{R})$ and $u^{(n)} \in L_1(\mathbb{R})$. Then*

$$\mathscr{F}(u^{(n)})(\xi) = (-i\xi)^n \mathscr{F}(u)(\xi).$$

Proof The detailed proof can be seen in [23, page 194]. See Exercise 2.3 for a related exercise. □

We would like to have a similar result for derivatives of fractional order.

The definition of the left Riemann-Liouvile fractional derivative suggests that for non-integer α, $\alpha \in (0, 1)$, an adequate class of functions [23, page 402] is

$$\{u \in L_1(\mathbb{R}) : I_+^{1-\alpha} u \in AC(\mathbb{R}),\ u' \in L_1(\mathbb{R})\}.$$

Similarly, for the right Riemann-Liouvile fractional derivative, consider

$$\{u \in L_1(\mathbb{R}) : I_-^{1-\alpha} u \in AC(\mathbb{R}),\ u' \in L_1(\mathbb{R})\}.$$

For other values of α, with $n - 1 < \alpha < n$, an adequate class of functions are respectively

$$\{u \in L_1(\mathbb{R}) : I_+^{n-\alpha} u \in AC^{(n-1)}(\mathbb{R}),\ u^{(n)} \in L_1(\mathbb{R})\},$$

$$\{u \in L_1(\mathbb{R}) : I_-^{n-\alpha} u \in AC^{(n-1)}(\mathbb{R}),\ u^{(n)} \in L_1(\mathbb{R})\}.$$

Proposition 2.3 *Let $n - 1 < \alpha < n$, $u \in L_1(\mathbb{R})$ and $u^{(n)} \in L_1(\mathbb{R})$.*

(a) *If $I_+^{n-\alpha} u \in AC^{(n-1)}(\mathbb{R})$, then*

$$\mathscr{F}(D_+^\alpha u)(\xi) = (-i\xi)^\alpha \mathscr{F}(u)(\xi).$$

(b) *If $I_-^{n-\alpha} u \in AC^{(n-1)}(\mathbb{R})$, then*

$$\mathscr{F}(D_-^\alpha u)(\xi) = (i\xi)^\alpha \mathscr{F}(u)(\xi).$$

Proof The proof of this proposition follows by direct calculation. We have by the definition and using the known property in the previous proposition

$$\mathscr{F}(D_+^\alpha u)(\xi) = \mathscr{F}\left(\frac{1}{\Gamma(n-\alpha)} \frac{d^n}{dx^n} \int_{-\infty}^x u(\eta)(x-\eta)^{n-1-\alpha} d\eta\right)$$

$$= \frac{1}{\Gamma(n-\alpha)}(-i\xi)^n \mathscr{F}\left(\int_{-\infty}^x u(\eta)(x-\eta)^{n-1-\alpha} d\eta\right).$$

By a change of variables and using the definition of Fourier transform and the theorem of Fubini it follows

$$\mathcal{F}(D_+^\alpha u)(\xi) = \frac{1}{\Gamma(n-\alpha)}(-i\xi)^n \frac{1}{\sqrt{2\pi}} \int_{\mathbb{R}} e^{ix\xi} \left(\int_0^\infty u(x-t)t^{n-1-\alpha} dt \right) dx$$

$$= \frac{1}{\Gamma(n-\alpha)}(-i\xi)^n \int_0^\infty t^{n-1-\alpha} \left(\frac{1}{\sqrt{2\pi}} \int_{\mathbb{R}} e^{ix\xi} u(x-t) dx \right) dt$$

$$= \frac{1}{\Gamma(n-\alpha)}(-i\xi)^n \int_0^\infty t^{n-1-\alpha} e^{it\xi} \mathcal{F}(u)(\xi) dt$$

$$= \frac{1}{\Gamma(n-\alpha)}(-i\xi)^n \mathcal{F}(u)(\xi) \int_0^\infty t^{n-1-\alpha} e^{it\xi} dt$$

$$= \frac{1}{\Gamma(n-\alpha)}(-i\xi)^n \mathcal{F}(u)(\xi) \xi^{-n+\alpha} \int_0^\infty t^{n-1-\alpha} e^{it} dt.$$

From the definition of the Gamma function it is easy to check that [4, page 36]

$$\int_0^\infty t^{x-1} e^{-ct} dt = \Gamma(x) c^{-x}$$

and therefore, for $n-1 < \alpha < n$, we have

$$\int_0^\infty t^{n-1-\alpha} e^{it} dt = \Gamma(n-\alpha)(-i)^{-n+\alpha}.$$

Finally, we arrive at

$$\mathcal{F}(D_+^\alpha u)(\xi) = (-i\xi)^\alpha \mathcal{F}(u)(\xi).$$

In a similar way we obtain

$$\mathcal{F}(D_-^\alpha u)(\xi) = (i\xi)^\alpha \mathcal{F}(u)(\xi).$$

$\square$

Additional discussions on the spaces of functions that can be considered for dealing with the Fourier transform of the Riemann-Liouville fractional derivatives can be found in [119, Section 7.1] and [23, Chapter 11]. In [119, page 139] it is considered the space of functions that are $C^{(n)}(\mathbb{R})$ and that vanish sufficiently rapidly at infinity together with their derivatives. The latter means that for each $k = 0, 1, 2, \ldots$ there exists m such that $\sup(1 + |x|)^m |u^{(k)}(x)| < \infty$. Other spaces can be considered such as the Lizorkin spaces. The characterisation of these spaces can be seen for instance in [119, page 148].

Many times the fractional derivatives are defined, in terms of the Fourier transform, as

$$D_+^\alpha u(x) = \mathscr{F}^{-1}((-i\xi)^\alpha \mathscr{F}(u))(x),$$

$$D_-^\alpha u(x) = \mathscr{F}^{-1}((i\xi)^\alpha \mathscr{F}(u))(x).$$

This is related to the physical interpretation that we will discuss in the next section. However, the operator is well defined only if $(\pm i\xi)^\alpha \mathscr{F}(u) \in L_1$. Therefore, the set of functions where this inversion is well defined, in the case of the left Riemann-Liouville derivative, is the set

$$\{u \in L_1(\mathbb{R}) : (-i\xi)^\alpha \mathscr{F}(u)(\xi) \in L_1(\mathbb{R})\}$$

and for the right Riemann-Liouville derivative, the appropriate set is

$$\{u \in L_1(\mathbb{R}) : (i\xi)^\alpha \mathscr{F}(u)(\xi) \in L_1(\mathbb{R})\}.$$

We can find in [23, Chapter 11] discussions on the caracterization of these spaces.

In the next result, we consider the conditions of Proposition 2.3 intersected with the previous set of functions.

Theorem 2.3 *Let* $n - 1 < \alpha < n$, $u \in L_1(\mathbb{R})$, $u^{(n)} \in L_1(\mathbb{R})$.

(a) If $I_+^{n-\alpha} u \in AC^{(n-1)}(\mathbb{R})$ *and* $(-i\xi)^\alpha \mathscr{F}(u)(\xi) \in L_1(\mathbb{R})$, *then the operator*

$$D_+^\alpha u(x) = \mathscr{F}^{-1}((-i\xi)^\alpha \mathscr{F}(u))(x)$$

is well defined for all $x \in \mathbb{R}$.

(b) If $I_-^{n-\alpha} u \in AC^{(n-1)}(\mathbb{R})$ *and* $(i\xi)^\alpha \mathscr{F}(u)(\xi) \in L_1(\mathbb{R})$, *then the operator*

$$D_-^\alpha u(x) = \mathscr{F}^{-1}((i\xi)^\alpha \mathscr{F}(u))(x)$$

is well defined for all $x \in \mathbb{R}$.

Proof This is a straightforward consequence of the fact that $(\pm i\xi)^\alpha \mathscr{F}(u)$ is in $L_1(\mathbb{R})$. $\qquad\square$

2.3 Lévy Flights and the Fractional Diffusion Equation

We start to recall the main ideas behind classical diffusion to establish a comparison to the anomalous diffusion described by the fractional diffusion equation.

Fig. 2.1 Evolution of
classical diffusion over time.
Plot of the function for two
instants of time

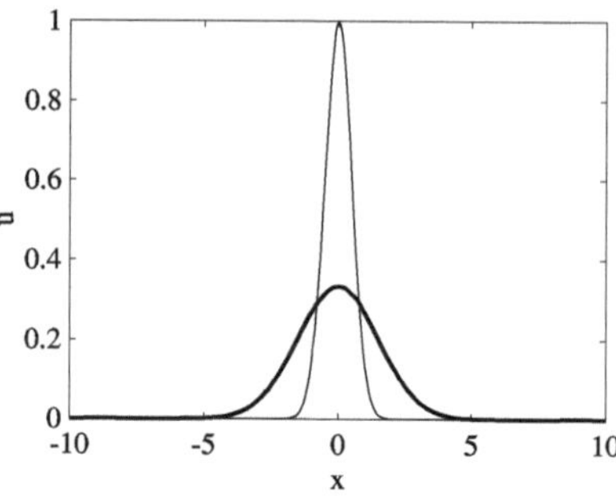

Classical diffusion is associated with Brownian motion [50, 112]. The density of
Brownian particles u at point x at time t, satisfies

$$\frac{\partial u}{\partial t}(x, t) = D\frac{\partial^2 u}{\partial x^2}(x, t).$$

The fundamental solution of this equation is

$$u(x, t) = \frac{1}{\sqrt{\pi 2Dt}}e^{-x^2/2Dt}, \quad x \in \mathbb{R}, \ t > 0.$$

This means that for a particle that starts from the origin at the initial time $t = 0$, the solution of the equation is the probability density function associated with
Gaussian (normal) distribution (see Fig. 2.1). A linear relation between the mean-
squared displacement x and t is given by

$$< x^2(t) > = \int_{\mathbb{R}} x^2 u(x, t)dx = 2Dt.$$

In classical diffusion the particle motion is described by the Brownian motion,
whereas in the type of anomalous diffusion that we will be discussing here, the
particle motion is described by Lévy flights. Lévy flights differ from Brownian
motion in the probability distribution of the jumps. For the Brownian motion the
increments are according to the Gaussian distribution and for the Lévy flights the
increments are according to a Lévy alpha stable distribution (see Fig. 2.2).

The probability distributions characterizing anomalous diffusive behaviour with
Lévy stable laws is nowadays a subject of an intense research activity on the physical
and mathematical communities [57, 69, 152]. To study these distributions, many
times a fractional differential equation is considered and its solutions are studied
[98, 100, 159, 160, 166].

Lévy flights stand for a class of non-Gaussian random processes whose stationary
increments are distributed according to a Lévy alpha-stable distribution. Lévy stable
laws are important for various reasons, one of them being the fact that the relevance

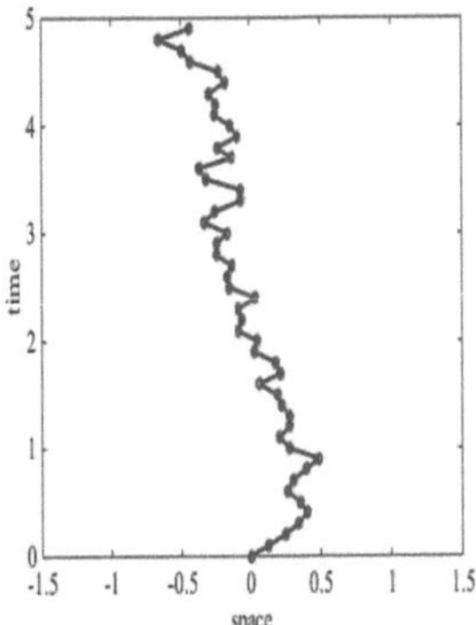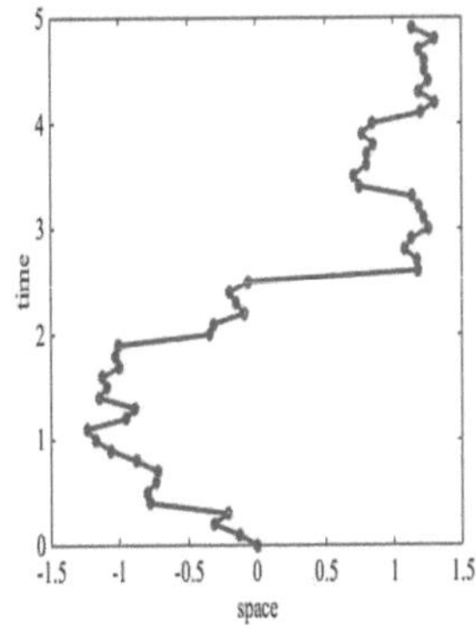

Fig. 2.2 Brownian motion (left) versus Lévy flights (right). Note the presence of large jumps in Lévy flights when compared to Brownian motion

of this motion has been observed in various scientific areas, where scale invariance phenomena takes place. We give a brief overview of Lévy flights underlying stable laws. Then, we arrive at the relation between these distributions and the fractional advection diffusion equation.

Lévy alpha-stable distributions are a four parameter family of distributions usually denoted by $S(\alpha, \beta, \gamma, \delta)$. The parameter α is called the characteristic exponent and describes the tail of the distribution. The skewness parameter $\beta \in [-1, 1]$ specifies if the distribution is right ($\beta > 0$) or left ($\beta < 0$) skewed, or symmetric ($\beta = 0$). The remaining parameters $\gamma > 0$ is the scale and $\delta \in \mathbb{R}$ is the location.

This class of stable distributions has a polynomial decay $|x|^{-1-\alpha}$ for $|x| \to \infty$, $\alpha \in (0, 2)$ and the mean square displacement for these distributions is infinite. The distribution tail increases with decreasing α. The distribution has a sharper peak and fatter tails.

To deal with theses distributions may present some difficulties related to the fact that they have infinite means and variances. Additionally, the probability density function or cumulative distribution function usually can not be described explicitly. For that reason, many results are obtained through the characteristic function $\psi(\xi)$ linked to the probability density function by the Fourier transform

$$\psi(\xi) = \frac{1}{\sqrt{2\pi}} \int_{\mathbb{R}} e^{i\xi x} u(x, t) dx.$$

One class of Lévy stable processes is the class for which the Fourier transform of the jump distribution [17, 105, 120], for $\alpha \in (0, 2)$, $\alpha \neq 1$, is given by

$$\psi(\xi) = \exp\left[i\xi\delta - \gamma^{\alpha}|\xi|^{\alpha}(1 - i\beta\mathrm{sign}(\xi)\tan(\alpha\pi/2))\right]. \tag{2.5}$$

According to [118], the probability density function is positive if we have $\alpha \in (0, 2]$. The case $\alpha = 2$ corresponds to the Gaussian case, while $\alpha \in (0, 2)$ describes Lévy flights where very large jumps will typically occur.

We are interested in the class of Lévy processes described by the characteristic function, for $\alpha \in (1, 2]$, and where the δ and γ values are given respectively by

$$\delta = Vt,$$

$$\gamma^{\alpha} = |\cos(\pi\alpha/2)|Dt.$$

Hence,

$$\psi(\xi) = \exp\left[i\xi Vt - |\cos(\pi\alpha/2)|Dt|\xi|^{\alpha}(1 - i\beta\operatorname{sign}(\xi)\tan(\alpha\pi/2))\right].$$

Here, t represents the time, V and D are positive constants. The location involves the constant V and will be related to the advective coefficient of the fractional advection diffusion equation. The advection is the transport of a substance which in general is a fluid. The scale D will be related to the diffusive coefficient of that equation.

Note that

$$\frac{1+\beta}{2}(-i\xi)^{\alpha} + \frac{1-\beta}{2}(i\xi)^{\alpha} = |\xi|^{\alpha}\cos(\alpha\pi/2)(1 - i\beta\operatorname{sign}(\xi)\tan(\alpha\pi/2)).$$

Therefore, the above characteristic function can now be written for the new parameters γ and δ as

$$\psi(\xi) = \exp\left[i\xi Vt + Dt\left(\frac{1+\beta}{2}(-i\xi)^{\alpha} + \frac{1-\beta}{2}(i\xi)^{\alpha}\right)\right].$$

The previous characteristic function $\psi(\xi)$ represents the Fourier transform of u. Denote it as $\mathscr{F}(u)(\xi, t)$. Then, we note that the characteristic function is the solution of the following differential equation [98, 166]

$$\frac{\partial \mathscr{F}(u)(\xi, t)}{\partial t} = \left(i\xi V + D\left(\frac{1+\beta}{2}(-i\xi)^{\alpha} + \frac{1-\beta}{2}(i\xi)^{\alpha}\right)\right)\mathscr{F}(u)(\xi, t).$$

$$(2.6)$$

By applying the inverse Fourier transform we rewrite

$$\frac{\partial u(x, t)}{\partial t} = V\mathscr{F}^{-1}[i\xi\mathscr{F}(u)(\xi, t)] + D\frac{1+\beta}{2}\mathscr{F}^{-1}[(-i\xi)^{\alpha}\mathscr{F}(u)(\xi, t)]$$

$$+ D\frac{1-\beta}{2}\mathscr{F}^{-1}[(i\xi)^{\alpha}\mathscr{F}(u)(\xi, t)].$$

We recognise the inverse Fourier transforms from Sect. 2.2 and they represent fractional derivatives defined on the real line.

Let us denote these operators, that are defined in the real line for x, and also depend on t, in the following way

$$\frac{\partial^\alpha u}{\partial x^\alpha}(x, t) := \mathscr{F}^{-1}\left\{(-i\xi)^\alpha \mathscr{F}(u)(\xi, t)\right\}, \tag{2.7}$$

$$\frac{\partial^\alpha u}{\partial(-x)^\alpha}(x, t) := \mathscr{F}^{-1}\left\{(i\xi)^\alpha \mathscr{F}(u)(\xi, t)\right\}. \tag{2.8}$$

From Theorem 2.3, we infer that we have the left and right Riemann-Liouville fractional derivatives of order $\alpha \in (1, 2]$, that were previously defined. This means

$$\frac{\partial^\alpha u}{\partial x^\alpha}(x, t) = \frac{1}{\Gamma(2 - \alpha)} \frac{\partial^2}{\partial x^2} \int_{-\infty}^{x} u(\tau, t)(x - \tau)^{1-\alpha} d\tau \tag{2.9}$$

$$\frac{\partial^\alpha u}{\partial(-x)^\alpha}(x, t) = \frac{1}{\Gamma(2 - \alpha)} \frac{\partial^2}{\partial x^2} \int_{x}^{\infty} u(\tau, t)(\tau - x)^{1-\alpha} d\tau. \tag{2.10}$$

In this way we arrive at a final equation, for $\alpha \in (1, 2]$ and $\beta \in [-1, 1]$,

$$\frac{\partial u}{\partial t}(x, t) + V \frac{\partial u}{\partial x}(x, t) = D\nabla_\beta^\alpha u(x, t), \tag{2.11}$$

where,

$$\nabla_\beta^\alpha u(x, t) = \frac{1}{2}(1 + \beta)\frac{\partial^\alpha u}{\partial x^\alpha}(x, t) + \frac{1}{2}(1 - \beta)\frac{\partial^\alpha u}{\partial(-x)^\alpha}(x, t). \tag{2.12}$$

From what we have discussed in this section we can conclude that the alpha stable distributions play the role of fundamental solutions to the Eq. (2.11). Other interesting discussions about the relation of Lévy flights and diffusive processes can be found in [98, 159, 166].

The diffusion described by equations with fractional derivatives in space of order $\alpha \in (1, 2)$ is said to be an anomalous diffusion, known as superdiffusion. However, this is a generalisation of the usual definition of superdiffusion, since for the Lévy flights the variance of the process diverges. The supperdiffusive caracterization is usually associated with problems which second order moment scales as

$$< x^2(t) > \simeq t^\gamma, \quad \gamma > 1.$$

Therefore, the characterisation of the anomalous diffusion, in the case of Lévy flights, does not come from the behaviour of the second order moment but comes from the fractional moments given by

$$< |x(t)|^\eta > \simeq t^{\eta/\alpha}, \quad \text{with} \quad 0 < \eta < \alpha.$$

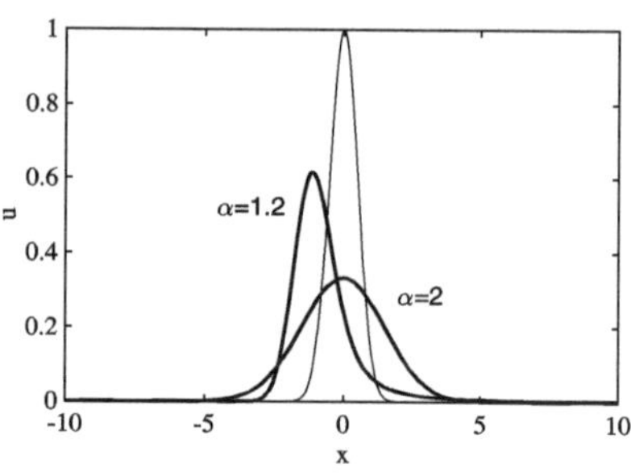

Fig. 2.3 Initial time represented by the thinner line. Evolution of diffusion over time, when $\beta = 1$, for $\alpha = 1.2$ and $\alpha = 2$

Then, to find the superdiffusive character [121], we can rescale this fractional moment according to

$$< |x(t)|^\eta >^{2/\eta} \simeq t^{2/\alpha}.$$

In the next chapter, we discuss how to find numerical solutions for the problem which consists in Eq. (2.11) defined in the whole real line, where V and D are positive constants, with initial condition

$$u(x, 0) = u_0(x), \quad x \in IR \tag{2.13}$$

and subject to the natural boundary conditions

$$\lim_{x \to -\infty} u(x, t) = 0 \quad \text{and} \quad \lim_{x \to \infty} u(x, t) = 0. \tag{2.14}$$

Observe in Fig. 2.3 the solution of Eq. (2.11), when $\beta = 1$, for $\alpha = 1.2$ and for the classical diffusion case $\alpha = 2$.

The access to the closed form of solutions related to this type of equations is very limited as can be seen in [20, 58, 92, 95, 101, 105, 107]. Nevertheless, in the next section, we present some of the analytical solutions we can find in the literature.

2.4 Exact Solutions for the Fractional Diffusion Equation

Consider the fractional diffusion equation, for $\alpha \in (1, 2]$,

$$\frac{\partial u}{\partial t}(x, t) = D \frac{\partial^\alpha u}{\partial |x|^\alpha}(x, t), \tag{2.15}$$

where the fractional operator on the right hand side is called the Riesz operator [98, 166] and defined through the Fourier transform as

$$\frac{\partial^{\alpha} u}{\partial |x|^{\alpha}}(x, t) := \mathscr{F}^{-1}\left[-|\xi|^{\alpha} \mathscr{F}(u)(\xi, t)\right]. \tag{2.16}$$

For this particular case we are able to obtain explicit analytical solutions.

The fractional diffusion equation (2.15) can be obtained taking in (2.5) respectively the values

$$\beta = 0, \quad \delta = 0, \quad \text{and} \quad \gamma^{\alpha} = Dt,$$

which leads to an equation similar to (2.6) of the form

$$\frac{\partial \mathscr{F}(u)(\xi, t)}{\partial t} = -D|\xi|^{\alpha} \mathscr{F}(u)(\xi, t). \tag{2.17}$$

By applying the inverse Fourier transform we obtain

$$\frac{\partial u(x, t)}{\partial t} = D\mathscr{F}^{-1}\left[-|\xi|^{\alpha} \mathscr{F}(u)(\xi, t)\right] \tag{2.18}$$

and therefore arrive at Eq. (2.15).

We can relate the Riesz operator with the Riemann-Liouville fractional derivatives. Since

$$\mathscr{F}\left(\frac{\partial^{\alpha} u}{\partial |x|^{\alpha}}\right)(\xi, t) = -|\xi|^{\alpha} \mathscr{F}(u)(\xi, t)$$

and (see Exercise 2.5)

$$|\xi|^{\alpha} = \frac{1}{2\cos(\alpha\pi/2)}\left((i\xi)^{\alpha} + (-i\xi)^{\alpha}\right)$$

we can obtain

$$-|\xi|^{\alpha} \mathscr{F}(u)(\xi, t) = \frac{-1}{2\cos(\alpha\pi/2)}\left((i\xi)^{\alpha} + (-i\xi)^{\alpha}\right)\mathscr{F}(u)(\xi, t).$$

Hence, the Riesz operator can be defined in terms of the Riemann-Liouville fractional derivatives, that is,

$$\frac{\partial^{\alpha} u}{\partial |x|^{\alpha}}(x, t) = \frac{-1}{2\cos(\alpha\pi/2)}\left(\frac{\partial^{\alpha} u}{\partial x^{\alpha}}(x, t) + \frac{\partial^{\alpha} u}{\partial(-x)^{\alpha}}(x, t)\right). \tag{2.19}$$

This can be seen as a particular case of (2.12), where $\beta = 0$, that is, the symmetric case. It is now easy to conclude that

$$\frac{\partial^\alpha u}{\partial |x|^\alpha}(x, t) = \frac{-1}{2\cos(\pi\alpha/2)}\frac{1}{\Gamma(2-\alpha)}\frac{\partial^2}{\partial x^2}\int_{-\infty}^{\infty} u(\tau, t)|\tau - x|^{1-\alpha}d\tau. \qquad (2.20)$$

An explicit form of the fundamental solution of Eq. (2.15) is given in terms of the Fox functions [98, 122], that is,

$$u_\alpha(x, t) = \frac{1}{\alpha|x|}H_{2,2}^{1,1}\left[\frac{|x|}{(Dt)^{1/\alpha}}\left|\begin{matrix}(1, 1/\alpha) & (1, 1/2)\\(1, 1) & (1, 1/2)\end{matrix}\right.\right]. \qquad (2.21)$$

A detailed discussion on the derivation of this solution and the relevant properties of the Fox functions in this context can be found in [93].

The expression (2.21) can be simplified [98, 141] resulting in

$$u_\alpha(x, t) = \frac{1}{\alpha\pi(Dt)^{1/\alpha}}\sum_{n=0}^{\infty}\frac{\Gamma((1+2n)/\alpha)}{\Gamma(1+2n)}(-1)^n\left(\frac{x^2}{(Dt)^{2/\alpha}}\right)^n. \qquad (2.22)$$

For $\alpha = 2$ we obtain the well-know normal distribution

$$u_2(x, t) = \frac{1}{\sqrt{4\pi Dt}}e^{-x^2/4Dt}.$$

The fundamental solutions are helpful to derive solutions for more general problems.

Denote by $K_\alpha(x, t)$ the fundamental solution of Eq. (2.15). The solution of Eq. (2.15) subjected to certain initial conditions,

$$u(x, 0) = u_0(x),$$

is representable by the convolution formula

$$u(x, t) = \int_{\mathbb{R}} u_0(\tau)K_\alpha(x - \tau, t)d\tau. \qquad (2.23)$$

To arrive at this formula we start with the fractional differential equation (2.15). We apply the Fourier transform to this equation, considering that the function is under the conditions we have discussed in Sect. 2.2, to obtain

$$\frac{\partial\mathscr{F}(u)(\xi, t)}{\partial t} = -D|\xi|^\alpha\mathscr{F}(u)(\xi, t).$$

Consequently, assuming that $\mathscr{F}(u_0) \in L_1(\mathbb{R})$,

$$\mathscr{F}(u)(\xi, t) = \mathscr{F}(u)(\xi, 0)\exp(-Dt|\xi|^{\alpha}).$$

Since $\exp(-Dt|\xi|^{\alpha})$ is continuous and bounded, the result follows taking the inverse Fourier transform. Using the convolution property of Fourier transforms, we obtain (2.23).

This explicit representation with a kernel allows that in this case solutions are C^{∞} and bounded for every $t > 0$ and $x \in \mathbb{R}$, under the assumption that the initial data is integrable [3, 16, 152]. For additional information on the explicit solution see also [105].

Remark 2.1 In [12] the result (2.23) is shown under the conditions $u \geq 0$, $u_0 \in C(\mathbb{R}) \cap L_{\infty}(\mathbb{R})$, $u_t \in C(\mathbb{R} \times (0, T))$ is continuous, $u \in C(\mathbb{R} \times [0, T))$, $\partial^{\alpha} u/\partial |x|^{\alpha} \in C(\mathbb{R} \times (0, T))$ and Eq. (2.15) is satisfied pointwise for every point $(x, t) \in \mathbb{R} \times (0, T)$. It is also mentioned that $e^{-|\xi|^{\alpha}t}$ is a tempered distribution and that $K_{\alpha} \in C^{\infty}(\mathbb{R} \times (0, \infty))$. Consequently in view of the previews result and the fact that the derivatives both in space and in time of K_{α} have a good decay at infinity, we will have $u \in C^{\infty}(\mathbb{R} \times (0, T))$.

Remark 2.2 For the asymmetric cases, $\beta \neq 0$, although the solution can have the representation (2.23), in general, we do not know the explicit expressions of the fundamental solutions $K_{\alpha}(x, t)$. One of the few known asymmetric distributions is the Lévy-Smirnov distribution [149].

Formula (2.23) gives an explicit expression of the solution of the fractional partial differential equation in terms of the initial data u_0. Because $K_{\alpha}(x, t) > 0$, for all $x \in \mathbb{R}$, $t > 0$, and

$$\int_{\mathbb{R}} K_{\alpha}(y, t)dy = 1, \quad \text{for all } t > 0,$$

we conclude that if u_0 is a bounded continuous function, then, for all $t > 0$,

$$\sup_{x \in \mathbb{R}} |u(x, t)| \leq \sup_{x \in \mathbb{R}} |u_0(x)|.$$

Hence,

$$\|u\|_{L_{\infty}(\mathbb{R})} \leq \|u_0\|_{L_{\infty}(\mathbb{R})}.$$

In order to obtain a similar result in the L_2 norm we give the Parseval's identity.

Theorem 2.4 (Parseval's Identity) *Let* $u, \mathscr{F}(u) \in L_1(\mathbb{R})$ *and* $u \in L_2(\mathbb{R})$. *Then* $\mathscr{F}(u) \in L_2(\mathbb{R})$ *and*

$$\|u\|_{L_2(\mathbb{R})} = \|\mathscr{F}(u)\|_{L_2(\mathbb{R})}.$$

Proof We begin by observing that

$$\int_{\mathbb{R}} \mathscr{F}(u)(\xi)v(\xi)d\xi = \int_{\mathbb{R}} \left(\frac{1}{\sqrt{2\pi}} \int_{\mathbb{R}} u(x)e^{ix\xi}dx \right) v(\xi)d\xi$$

$$= \int_{\mathbb{R}} \left(\frac{1}{\sqrt{2\pi}} \int_{\mathbb{R}} v(\xi)e^{ix\xi}d\xi \right) u(x)dx$$

$$= \int_{\mathbb{R}} u(x)\mathscr{F}(v)(x)dx.$$

We denote the conjugate of a complex number w by $\overline{w}$. Let us take $v(\xi) = \overline{\mathscr{F}(u)(\xi)}$. We have

$$v(\xi) = \overline{\mathscr{F}(u)(\xi)} = \overline{\frac{1}{\sqrt{2\pi}} \int_{\mathbb{R}} u(x)e^{ix\xi}dx}$$

$$= \frac{1}{\sqrt{2\pi}} \int_{\mathbb{R}} \overline{u}(x)e^{-ix\xi}dx$$

$$= \mathscr{F}^{-1}(\overline{u}).$$

Therefore,

$$\int_{\mathbb{R}} \mathscr{F}(u)(\xi)\overline{\mathscr{F}(u)(\xi)}d\xi = \int_{\mathbb{R}} u(x)\mathscr{F}(\mathscr{F}^{-1}(\overline{u}))(x)dx.$$

This is the same as

$$||\mathscr{F}(u)||_{L_2(\mathbb{R})} = ||u||_{L_2(\mathbb{R})}.$$

$\square$

We have, for each $t > 0$, that

$$||u(\cdot, t)||_{L_2(\mathbb{R})} = ||\mathscr{F}(u(\cdot, t)||_{L_2(\mathbb{R})}$$

and therefore for our equation we have

$$||u(\cdot, t)||_{L_2(\mathbb{R})} = ||e^{-Dt|k|^{\alpha}}\mathscr{F}(u_0)(\cdot, t)||_{L_2(\mathbb{R})}$$

$$\leq ||\mathscr{F}(u_0)(\cdot, t)||_{L_2(\mathbb{R})}$$

$$= ||u_0||_{L_2(\mathbb{R})}.$$

The solution depends continuouslly from the initial data in the L_∞ norm as seen previously and in the L_2 norm as we have now seen.

2.5   Exercises and Solutions

Exercises

2.1 Prove that fractional integrals are defined for functions in $L_1(\mathbb{R})$ existing almost everywhere.

2.2 Prove that if $f, g \in L_1(\mathbb{R})$, then $(f * g)(x)$ exist almost everywhere as an absolutely convergent integral, $f * g \in L_1(\mathbb{R})$ and

$$\|f * g\|_{L_1(\mathbb{R})} \le \|f\|_{L_1(\mathbb{R})} \|g\|_{L_1(\mathbb{R})}.$$

2.3 Show that if $u \in AC^{(n-1)}(\mathbb{R})$ and $u^{(r)} \in L_1(\mathbb{R})$, $r = 0, 1, 2, \ldots, n$, then

$$\mathscr{F}(u^{(n)})(\xi) = (-i\xi)^n \mathscr{F}(u)(\xi).$$

2.4 Let $u \in L_1(\mathbb{R})$, $I_-^{n-\alpha} u \in AC^{(n-1)}(\mathbb{R})$ and $u^{(n)} \in L_1(\mathbb{R})$. Prove that

$$\mathscr{F}(D_-^{\alpha} u)(\xi) = (i\xi)^{\alpha} \mathscr{F}(u)(\xi).$$

2.5 Show that, for $\alpha > 0$,

$$|\xi|^{\alpha} = \frac{1}{2\cos(\alpha\pi/2)}((i\xi)^{\alpha} + (-i\xi)^{\alpha}).$$

2.6 (a) Let A be a positive constant and u be the function defined in $\mathbb{R}$ by $u(x) = e^{-Ax^2}$. Prove that

$$\mathscr{F}(u)(\xi) = \frac{1}{\sqrt{2A}} e^{-\xi^2/4A}.$$

(b) Use (a) to determine the solution of the diffusion problem, that is, the case $\alpha = 2$,

$$\frac{\partial u}{\partial t} = D \frac{\partial^2 u}{\partial x^2}$$

with $u(x, 0) = u_0(x)$.

2.7 Consider the fractional diffusion equation, for $\alpha \in (1, 2]$,

$$\frac{\partial u}{\partial t}(x, t) = D \frac{\partial^{\alpha} u}{\partial |x|^{\alpha}}(x, t) - ru(x, t) + ru_0(x), \tag{2.24}$$

where the fractional operator on the right hand side is the Riesz operator, $r > 0$ represents a resetting property, $u_0(x)$ is the initial condition and we also have

$$\lim_{|x| \to \infty} u(x, t) = 0.$$

Using Fourier transforms prove that the solution of this problem is given by

$$u(x, t) = e^{-rt}u_\alpha(x, t) + \int_0^t e^{-r(t-s)}u_\alpha(x, t - s)ds,$$

where $u_\alpha(x, t)$ is the solution of the problem when $r = 0$.

2.8 For $\sigma \in (0, 1)$, the fractional Laplacian of order σ is defined as

$$\mathscr{F}[(-\Delta)^\sigma u](\xi) = |\xi|^{2\sigma}\mathscr{F}(u)(\xi).$$

(a) Show that, if $u \in \mathscr{S}$, where $\mathscr{S}$ is the Schwarz space, then the fractional Laplacian can be expressed by the pointwise formula

$$(-\Delta)^\sigma u(x) = \frac{4^\sigma\, \Gamma(1/2 + \sigma)}{\sqrt{\pi}\Gamma(-\sigma)} \int_\mathbb{R} \frac{u(x) - u(z)}{|x - z|^{1+2\sigma}} dz.$$

The Cauchy principal value is applied to compute the integral.

(b) Show that, if $u \in C^2(\mathbb{R}) \cap L_\infty(\mathbb{R})$ then

$$\lim_{\sigma \to 1^-} (-\Delta)^\sigma u(x) = -\Delta u(x),$$

where Δu is the Laplacian operator. In the one dimensional case it is the second order derivative.

(c) Show that, for $\sigma = \alpha/2$ we have

$$\frac{4^\sigma\, \Gamma(1/2 + \sigma)}{\sqrt{\pi}\Gamma(-\sigma)} = -\frac{2^{\alpha-1}\alpha\Gamma((1 + \alpha)/2)}{\sqrt{\pi}\Gamma(1 - \alpha/2)}.$$

(d) Show the fractional Laplacian is equal to the Riesz operator, that is

$$(-\Delta)^{\alpha/2}u(x) = -\frac{1}{2\cos(\alpha\pi/2)}(D_-^\alpha u(x) + D_+^\alpha u(x)).$$

2.9 Let $\alpha > 0$, $n = [\alpha] + 1$ and $u(x) \in AC^n(\mathbb{R})$ and define the Caputo fractional derivatives as

$$_cD_+^\alpha u(x) = \frac{1}{\Gamma(n - \alpha)} \int_{-\infty}^x u^{(n)}(\xi)(x - \xi)^{n-1-\alpha}d\xi.$$

$$_cD_-^\alpha u(x) = \frac{1}{\Gamma(n - \alpha)} \int_x^\infty u^{(n)}(\xi)(\xi - x)^{n-1-\alpha}d\xi.$$

(a) Prove that if

$$\lim_{a \to -\infty} (x - a)^{n-\alpha} u^{(k)}(a) = 0, \quad k = 0, 1, \ldots, n - 1,$$

then

$$_c D_+^\alpha u(x) = D_+^\alpha u(x).$$

(b) Prove that if

$$\lim_{b \to -\infty} (b - x)^{n-\alpha} u^{(k)}(b) = 0, \quad k = 0, 1, \ldots, n - 1,$$

then

$$_c D_-^\alpha u(x) = D_-^\alpha u(x).$$

Solutions

2.1 To prove the statement that fractional integrals are defined for functions in $L_1(\mathbb{R})$ existing almost everywhere, we need to divide the integral in the sum of two integrals [119, page 94], that is, we have

$$
\begin{aligned}
I_+^\alpha u(x) &= \frac{1}{\Gamma(\alpha)} \int_{-\infty}^{x} u(t)(x - t)^{\alpha-1} dt \\
&= \frac{1}{\Gamma(\alpha)} \int_{0}^{\infty} u(x - t) t^{\alpha-1} dt \\
&= \frac{1}{\Gamma(\alpha)} \int_{0}^{1} u(x - t) t^{\alpha-1} dt + \frac{1}{\Gamma(\alpha)} \int_{1}^{\infty} u(x - t) t^{\alpha-1} dt.
\end{aligned}
$$

The first integral exists, for almost every x, using the generalised Minkowsky inequality. The second integral exists, for all x, by Hölder's inequality, since $u \in L_1(1, \infty)$ and the function $t^{\alpha-1} \in L_\infty(1, \infty)$.

The Hölder inequality [119, page 8]: Let $f \in L_p(\Omega)$, $1 \le p \le \infty$ and $g \in L_{p'}(\Omega)$. Then $fg \in L_1(\Omega)$ and

$$\int_{\Omega} |f(x)g(x)| dx \le \|f\|_p \|g\|_{p'}, \quad p' = \frac{p}{p - 1}.$$

If p is such that $1 \le p \le \infty$ the conjugate number p' is defined through $1/p + 1/p' = 1$ in case $1 < p < \infty$. We have $p' = \infty$ if $p = 1$ and $p' = 1$ if $p = \infty$.

The generalised Minkowsky inequality [119, page 9]: Let $\Omega_1 = (a, b)$ and $\Omega_2 = (c, d)$, $-\infty \le a < b \le \infty$, $-\infty \le c < d \le \infty$, and let $f(x, y)$ be a measurable function defined on $\Omega_1 \times \Omega_2$. If at least one of the integrals

$$\int_{\Omega_1} dx \int_{\Omega_2} f(x, y) dy, \quad \int_{\Omega_2} dy \int_{\Omega_1} f(x, y) dx, \quad \int_{\Omega_1 \times \Omega_2} f(x, y) dx dy$$

is absolutely convergent then they coincide and we have

$$\left\{ \int_{\Omega_1} dx \left| \int_{\Omega_2} f(x, y) dy \right|^p \right\}^{1/p} \le \int_{\Omega_2} dy \left\{ \int_{\Omega_1} |f(x, y)|^p dx \right\}^{1/p}.$$

2.2 Start to see that

$$\frac{1}{\sqrt{2\pi}} \int_{\mathbb{R}} |f(x - \tau)||g(\tau)| dx = |g(\tau)| \|f\|_{L_1(\mathbb{R})}$$

which belongs to $L_1(\mathbb{R})$. It follows that

$$\frac{1}{\sqrt{2\pi}} \int_{\mathbb{R}} d\tau \left\{ \frac{1}{\sqrt{2\pi}} \int_{\mathbb{R}} |f(x - \tau)||g(\tau)| dx \right\} = \|f\|_{L_1(\mathbb{R})} \|g\|_{L_1(\mathbb{R})}$$

exists as a finite number. Therefore by Fubini's theorem we can change the order of integration and use it to conclude that

$$\frac{1}{\sqrt{2\pi}} \int_{\mathbb{R}} |f(x - \tau)||g(\tau)| d\tau$$

exists for almost all $x \in \mathbb{R}$ and belongs to $L_1(\mathbb{R})$. See [23, page 5] for a detailed proof of a more general case, that is, when $f \in L_p(\mathbb{R})$, $1 \le p < \infty$.

2.3 Start with the definition of Fourier transform for $u^{(n)}$, that is, $\mathscr{F}(u^{(n)})(\xi)$. Then arrive at an expression upon integration by parts. For the case $n = 1$ we would obtain $\mathscr{F}(u')(\xi) = (i\xi)\mathscr{F}(u)(\xi)$. After the integration by parts, the only thing we need to prove is that the limite of $u(x)$ vanishes as $|x| \to \infty$. Because the function is absolutely continuous in any finite interval we have

$$u(x) = u(0) + \int_0^x u'(t) dt$$

and the condition $u' \in L_1(\mathbb{R})$ implies that the limit exists as x goes to $\pm \infty$. Since $u \in L_1(\mathbb{R})$ we must have that the limit is zero.

2.4 The proof is similar to the proof presented in Proposition 2.3.

2.5 To arrive at the equality, it is enough to notice that

$$(\pm i\xi)^\alpha = |\xi|^\alpha \exp(\pm i\alpha\pi \operatorname{sgn}(\xi)/2).$$

2.6 (a) This is computed by direct calculations. Start by writing the definition of Fourier transform of u. Then by completing the square of the exponent of the exponential and by doing some change of variables arrive at an integral in the real line of the function e^{-x^2} which is equal to $\sqrt{\pi}$.

(b) To determine the solution of the classical diffusion equation, we apply the Fourier transform to the differential equation and arrive at

$$\frac{\partial \mathscr{F}(u)}{\partial t}(\xi, t) + |\xi|^2 \mathscr{F}(u)(\xi) = 0 \quad \text{in } \mathbb{R} \times (0, \infty)$$

and

$$\mathscr{F}(u)(\xi, 0) = \mathscr{F}(u_0)(\xi), \text{ on } \mathbb{R}.$$

This is an initial value problem for the ordinary differential problem with $\xi \in \mathbb{R}$ as a parameter. The solution is given by

$$\mathscr{F}(u)(\xi, t) = \mathscr{F}(u_0)(\xi)e^{-|\xi|^2 t},$$

treating t as a parameter. For any $t > 0$, let $K_2(x, t)$ satisfy

$$\mathscr{F}(K_2)(\xi, t) = e^{-|\xi|^2 t}.$$

Then

$$\mathscr{F}(u)(\xi, t) = \mathscr{F}(K_2)(\xi, t)\mathscr{F}(u_0)(\xi).$$

By the convolution property and (a) we arrive at

$$u(x, t) = \int_{\mathbb{R}} K_2(x - y, t)u_0(y)dy,$$

with

$$K_2(x, t) = \frac{1}{\sqrt{4\pi t}}e^{-x^2/4t}.$$

for any $(x, t) \in \mathbb{R} \times (0, \infty)$.

2.7 Just apply the Fourier transform in x and solve the ordinary differential equation in time.

2.8 (a) The proof can be found in [142]. In particular Lemma 5.1 in [142].

(b) The proof can be found in [142]. In particular Proposition 5.3 in [142].

(c) This is a direct calculation using the property $\Gamma(z + 1) = z\Gamma(z)$.

(d) This is a direct calculation using integration by parts. The constant can be manipulated using the properties of the Gamma function $\Gamma(z+1) = z\Gamma(z)$,

$\Gamma(2z) = \sqrt{\pi}2^{2z-1}\Gamma(z)\Gamma(z+1/2)$ and $\Gamma(z)\Gamma(1-z) = \pi/\sin(\pi z)$ to obtain

$$\frac{2^{\alpha-1}\alpha\Gamma((1+\alpha)/2)}{\sqrt{\pi}\,\Gamma(1-\alpha/2)} = \frac{1}{2\cos(\alpha\pi/2)}\frac{\alpha(1-\alpha)}{\Gamma(2-\alpha)}.$$

2.9 The proof is similar to the proof of Theorem 1.14 in Chap. 1.

Chapter 3
Consistency, Stability and Convergence of Numerical Methods

In this chapter, we define the concepts of consistency, stability and convergence of finite difference approximations for fractional partial differential equations. For the stability analysis, we introduce the von Neumann analysis, the eigenvalue analysis and the energy method. For the convergence of the numerical methods we present the Lax theorem and the strategy based on the energy method.

3.1 Introduction

The definitions presented in this section can be found in classical books on finite difference methods for classical partial differential equations, such as, [102, 103, 146]. We extend the concepts here to the context of finite difference methods for fractional partial differential equations.

Let

$$\mathscr{L}u = f$$

be a fractional partial differential equation, where u is a function that depends on $x \in \mathbb{R}$ and $t > 0$, f is a source term, and with an initial condition $u(x, 0) = u_0(x)$. Assume that we have obtained an approximate solution, from a difference scheme, that we shall refer to L_j^n, where n corresponds to the time step and j to the spatial mesh point. Let U_j^n be the approximate solution of $u(x_j, t_n)$ at the grid points defined by

$$x_j = j\Delta x, \quad j \in \mathbb{Z}, \quad t_n = n\Delta t, \quad n \geq 0.$$

© The Author(s), under exclusive license to Springer Nature Switzerland AG 2026
E. Sousa, *Finite Difference Methods for Fractional Diffusion Equations*, Lecture Notes in Mathematics 2389, https://doi.org/10.1007/978-3-032-11222-4_3

The finite difference scheme that approximates $\mathcal{L}u = f$ can be represented by

$$L_j^n U_j^n = f_j^n, \quad j \in \mathbb{Z}. \tag{3.1}$$

Assume that the difference operator L_j^n can be written in the following manner

$$L_j^n U_j^n = \frac{1}{\Delta t}(U_j^{n+1} - QU_j^n), \quad j \in \mathbb{Z}, \tag{3.2}$$

with Q an operator acting on U_j^n as

$$Q = \sum_{k=-r}^{p} a_k E^k, \quad E^a U_j^n = U_{j+a}^n, \tag{3.3}$$

where r and p can be infinity. The a_k's depend on the parameter Δx.

The finite difference method is defined by

$$U_j^{n+1} = QU_j^n + \Delta t f_j^n. \tag{3.4}$$

The vectorial form of this numerical method can be represented as

$$U^{n+1} = QU^n + \Delta t F^n \tag{3.5}$$

where $U^n = (\ldots, U_{-1}^n, U_0^n, U_1^n, \ldots)^T$, $F^n = (\ldots, f_{-1}^n, f_0^n, f_1^n, \ldots)^T$ and the operator Q acts in each of the components of the vector U^n according to (3.3).

Remark 3.1 The previous approach seems only suitable for explicit schemes. However, if we have an implicit scheme that uses two time steps, we can easily adjust the theory we are going to present to the implicit case. In the case of a two-step implicit method, equality (3.5) becomes

$$Q_1 U^{n+1} = QU^n + \Delta t F^{n+1}, \tag{3.6}$$

where $\Delta t F^{n+1}$ can depend on $n + 1$ or on both n and $n + 1$.

3.2 Consistency, Stability and von Neumann Analysis

A finite difference method depends on the space step Δx and time step Δt of the mesh. The numerical solution of the finite difference method should converge to the exact solution of the problem. This means that when Δx and Δt are small and goes to zero, the numerical solution should be a good approximation of the exact solution. Many times we measure the quality of the numerical solution using the concepts of consistency and stability. The consistency is related with the fact that the finite difference formulas in the limit, when Δx and Δt goes to zero, lead to the differential problem. Stability is related with the computational performance of the numerical method or how sensitive is the solution to the data we introduce. This is quite similar to the definition of well-posedness of a differential problem, where small perturbations in the initial data must lead to small perturbations in the final solution.

Definition 3.1 (Consistency) The difference scheme (3.5) is consistent with the fractional partial differential equation in a norm $|| \cdot ||$ if the solution of the fractional partial differential equation, u, satisfies,

$$u^{n+1} = Qu^n + \Delta t f^n + \Delta t T^n,$$

where u^n, f^n, T^n denotes vectors whose components are $u(x_j, t_n)$, $f(x_j, t_n)$, $T(x_j, t_n)$ respectively and $||T^n||$ goes to zero as $\Delta x, \Delta t$ goes to zero.

A slight variation of the definition of consistency is given bellow where the order of accuracy in which T^n goes to zero (that is the order of accuracy to which the finite difference scheme approximates the fractional partial differential equation) is included.

Definition 3.2 The finite difference scheme (3.5) is said to be accurate of order (p, q) to the fractional partial differential equation if

$$||T^n|| = \mathcal{O}((\Delta x)^p) + \mathcal{O}((\Delta t)^q).$$

We refer to T^n or $||T^n||$ as the truncation error.

The notation of big $\mathcal{O}$, used along this work, stands for the usual notation, that is, a function f satisfies $f(\Delta x) = \mathcal{O}((\Delta x)^p)$, $p \in \mathbb{R}$, if and only if there exists $M > 0$ and $\Delta x_0 > 0$ such that $|f(\Delta x)| \leq M(\Delta x)^p$, for all $\Delta x < \Delta x_0$.

Remark 3.2 In the definition of consistency Δx and Δt goes to zero arbitrarily. However, in certain cases, achieving consistency, or obtaining an higher order accuracy, may require imposing a specific relation between Δx and Δt.

Remark 3.3 In the case of a two step implicit scheme like (3.6) the consistency definition becomes

$$Q_1 u^{n+1} = Qu^n + \Delta t f^{n+1} + \Delta t r^n.$$

Hence, we can rewrite it as

$$u^{n+1} = Q_1^{-1} Qu^n + \Delta t Q_1^{-1} f^{n+1} + \Delta t Q_1^{-1} r^n$$

and consider $T^n = Q_1^{-1} r^n$. If $||Q_1^{-1}||$ is uniformly bounded as Δx and Δt approach zero, the consistency of the scheme with respect to the norm $|| \cdot ||$ is determined by r^n since $||T^n|| \leq ||Q_1^{-1}|| \, ||r^n||$.

For the particular case of the fractional diffusion equation, in the previous chapter, we have presented some of its exact solutions. We can find a stability property that replicates the property we have seen regarding the L_∞ and L_2 norms, that is, we have seen that in both norms $||u|| \leq ||u_0||$, where u is the exact solution of the fractional partial differential equation, given the initial condition u_0.

Let u_j^n be the exact solutions of the discretized equation (3.4) and $\overline{u}_j^n$ the actual computed solutions. The difference between the two solutions can be related to round-off errors and to errors in the initial data. Hence, if we define the error as the difference between these two solutions it is easy to check that the error satisfies the finite difference equation with no source term, that is,

$$U_j^{n+1} = Q U_j^n. \tag{3.7}$$

Definition 3.3 Let $\{U_j^n, j \in \mathbb{Z}, \}, n = 1, \ldots, N$, be a solution of (3.7). The finite difference scheme (3.4) is **practically** stable in the discrete l_2 norm if

$$||U^n||_{l_2} \leq ||U^0||_{l_2}, \quad n = 1, \ldots, N, \tag{3.8}$$

where

$$||U^n||_{l_2} = \left(\Delta x \sum_{j=-\infty}^{\infty} |U_j^n|^2 \right)^{1/2}.$$

The definition of stability postulate that the difference scheme should not allow errors to grow indefinitely to avoid the amplification of errors without a bound as we progress from one time step to another. The stability condition is a requirement on the numerical method and contains no condition on the differential equation.

To prove the stability in the l_2 norm we can use Fourier analysis. We start to define the semidiscrete Fourier transform.

Definition 3.4 The semidiscrete Fourier transform of a function U defined on the infinite mesh $x_j = j\Delta x$, $j = 0, \pm 1, \pm 2, \ldots$ is

$$\hat{U}(\xi) = \frac{1}{\sqrt{2\pi}} \Delta x \sum_{j=-\infty}^{\infty} U_j e^{-i\xi x_j}, \quad \xi \in \left[-\frac{\pi}{\Delta x}, \frac{\pi}{\Delta x} \right].$$

We also need the definition of the inverse semidiscrete Fourier transform, similarly to the case of the Fourier transform and its inverse considered earlier.

Definition 3.5 Let $\hat{U}$ be defined on the interval $[-\pi/\Delta x, \pi/\Delta x]$. The inverse semidiscrete Fourier transform of $\hat{U}$ is defined, for $x_j = j\Delta x$, by

$$U_j = \frac{1}{\sqrt{2\pi}} \int_{-\frac{\pi}{\Delta x}}^{\frac{\pi}{\Delta x}} \hat{U}(\xi)e^{i\xi x_j} d\xi.$$

We proceed with the presentation of the discrete Parseval's identity, necessary to prove some of the next results.

Theorem 3.1 (Discrete Parseval's Identity) *Let*

$$\|U\|_{l_2} = \left(\Delta x \sum_{j=-\infty}^{\infty} |U_j|^2 \right)^{1/2} \quad and \quad \|\hat{U}\|_{L_2} = \left(\int_{-\pi/\Delta x}^{\pi/\Delta x} |\hat{U}(\xi)|^2 d\xi \right)^{1/2}.$$

If $\|U\|_{l_2}$ *is finite, then also* $\|\hat{U}\|_{L_2}$ *is finite and*

$$\|U\|_{l_2} = \|\hat{U}\|_{L_2}.$$

Proof Similar to the proof of Parseval's identity in Chap. 2 and therefore this is left as an exercise at the end of the chapter. $\qquad\square$

In certain situations, the practical stability $\|U^n\|_{l_2} \leq \|U^0\|_{l_2}$ is too restrictive. Therefore, we present a more general definition and many times a more adequate definition for evolutionary problems.

Definition 3.6 Let $\{U_j^n, j \in \mathbb{Z}\}$, $n = 1, \ldots, N$, be a solution of (3.7). The finite difference scheme (3.4), on the time interval $[0, T]$, is **von Neumann stable** in the l_2 norm, if there exists a positive constant $C = C(T)$ such that

$$\|U^n\|_{l_2} \leq C\|U^0\|_{l_2}, \quad n = 1, \ldots, N = \frac{T}{\Delta t},$$

where

$$\|U^n\|_{l_2} = \left(\sum_{j=-\infty}^{\infty} \Delta x |U_j^n|^2 \right)^{1/2}.$$

Practical stability is when we have $C = 1$.

Since the **stability constant** usually depends on T and $C(T) \to \infty$ when $T \to \infty$ most of the times the von Neumann stability holds for finite time intervals $[0, T]$ $(T < \infty)$ and only for a limited range of $0 \leq n \leq T/\Delta t$. Von Neumann stability of a finite difference scheme can be easily verified by using the following result.

Lemma 3.1 *Suppose that the semidiscrete Fourier transform of the solution $\{U_j^n : j \in \mathbb{Z}\}$, $n = 0, 1, \ldots, T/\Delta t$ of a finite difference scheme satisfies*

$$\hat{U}^{n+1}(\xi) = \kappa(\xi)\hat{U}^n(\xi)$$

and

$$|\kappa(\xi)| \leq 1 + C_0\Delta t, \quad \text{for all } \xi \in \left[-\frac{\pi}{\Delta x}, \frac{\pi}{\Delta x}\right].$$

Then, the scheme is von Neumann stable. In particular, if $C_0 = 0$ then the scheme is practically stable.

Proof By the Parseval's identity for the semidiscrete Fourier transform and by the hypothesis we have the following

$$\begin{aligned}
||U^{n+1}||_{l_2} &= ||\hat{U}^{n+1}||_{L_2} \\
&= ||\kappa\hat{U}^n||_{L_2} \\
&\leq \max_\xi |\kappa(\xi)| \, ||\hat{U}^n||_{L_2} \\
&= \max_\xi |\kappa(\xi)| \, ||U^n||_{l_2}.
\end{aligned}$$

Hence,

$$||U^{n+1}||_{l_2} \leq (1 + C_0\Delta t)||U^n||_{l_2}, \quad n = 0, 1, \ldots, N - 1.$$

Therefore,

$$||U^n||_{l_2} \leq (1 + C_0\Delta t)^n||U^0||_{l_2}, \ n = 1, \ldots, N.$$

As $1 + C_0\Delta t \leq e^{C_0\Delta t}$ and

$$(1 + C_0\Delta t)^n \leq e^{C_0 n\Delta t} = e^{C_0 T},$$

it follows

$$||U^n||_{l_2} \leq C||U^0||_{l_2}, \ n = 1, \ldots, N,$$

with $C = e^{C_0 T}$. $\qquad\square$

We end this section with another definition of stability which is equivalent to the previous definition and now motivated by the last proof.

Definition 3.7 (Stability) The finite difference scheme

$$U^{n+1} = QU^n, \quad n \geq 0,$$

is said to be stable with respect to norm $|| \cdot ||$ if there exist positive constants Δx_0 and Δt_0 and, $k, \beta \geq 0$ so that

$$||U^n|| \leq k e^{\beta T} ||U^0||,$$

for $0 \leq T = n\Delta t$, $0 < \Delta x \leq \Delta x_0$ and $0 < \Delta t \leq \Delta t_0$.

The definitions of consistency and stability presented here are valid for finite difference schemes that are not defined in the whole real line. The stability analysis based on the Fourier analysis is appropriate for finite difference schemes defined in the real line. However, the conditions obtained through the von Neumann analysis, in the cases we have a bounded domain are in general necessary conditions for the stability of the numerical method.

3.3   Convergence and Lax Theorem

We proceed to discuss the convergence of a finite difference scheme. We start with some definitions and then present the Lax equivalence theorem.

Definition 3.8 (Convergence) A difference scheme $L_j^n U_j^n = f_j^n$ approximating the fractional partial differential equation $\mathscr{L}u = f$ is a convergent scheme at the time t if, as $(n+1)\Delta t \to t$, $||U^{n+1} - u^{n+1}|| \to 0$ as $\Delta x \to 0$ and $\Delta t \to 0$.

To have a measure of how fast is the convergence of a difference scheme we use the concept of order of convergence.

Definition 3.9 (Order of Convergence) A difference scheme $L_j^n U_j^n = f_j^n$ approximating the fractional partial differential equation $\mathscr{L}u = f$ is a convergent scheme of order (p, q) if for any t as $(n+1)\Delta t$ converges to t, $||U^{n+1} - u^{n+1}|| = \mathcal{O}((\Delta x)^p) + \mathcal{O}((\Delta t)^q)$ as Δx and Δt converge to 0.

The Lax equivalence theorem connects convergence, consistency and stability.

Theorem 3.2 (Lax Equivalence Theorem) *A consistent, two level finite difference scheme for a well-posed linear initial-value problem is convergent if and only if it is stable.*

The proof of the Lax Equivalence Theorem can be found in the classical book [113, pages 45–47] or in the paper of Lax and Richtmyer [79]. The definition of a properly posed problem can also be found in [113, pages 39–42] and it is according to Hadamard definition, that is, the existing solution is unique and continuously dependent on the initial data.

We prove a slightly stronger version of half of the theorem above. But before presenting the proof we need the following proposition.

Proposition 3.1 *The finite difference scheme*

$$U^{n+1} = QU^n, \ n \geq 0,$$

is stable with respect to the norm $|| \cdot ||$ *if and only if there exist positive constants* Δx_0 *and* Δt_0 *and non-negative constants* K *and* β *so that*

$$||Q^{n+1}|| \leq Ke^{\beta t},$$

for $0 \leq t = (n+1)\Delta t, \ 0 < \Delta x \leq \Delta x_0$ *and* $0 < \Delta t \leq \Delta t_0$.

Proof Since

$$U^{n+1} = QU^n = Q(QU^{n-1}) = Q^2 U^{n-1} = \cdots = Q^{n+1}U^0,$$

the definition of stability

$$||U^{n+1}|| \leq Ke^{\beta t}||U^0||$$

can be written as

$$||Q^{n+1}U^0|| \leq Ke^{\beta t}||U^0||$$

or

$$\frac{||Q^{n+1}U^0||}{||U^0||} \leq Ke^{\beta t}.$$

If the scheme is **stable** then

$$\frac{||Q^{n+1}U^0||}{||U^0||} \leq Ke^{\beta t}.$$

By taking the supremum over both sides over all non-zero vectors U^0, we get

$$||Q^{n+1}|| \leq Ke^{\beta t}.$$

We have used the definition of norm of an operator.

To perform the proof in the other direction, we start to assume that

$$||Q^{n+1}|| \leq Ke^{\beta t}.$$

Since $||Q^{n+1}U^0|| \leq ||Q^{n+1}|| \ ||U^0||$, then

$$||Q^{n+1}U^0|| \leq Ke^{\beta t}||U^0||.$$

This implies

$$||U^{n+1}|| \leq K e^{\beta t} ||U^0||,$$

the definition of stability. $\square$

The proof of the next theorem uses the previous result.

Theorem 3.3 (Lax Theorem) *If a two level linear finite difference scheme*

$$U^{n+1} = QU^n$$

is accurate of order (p, q) in the norm $|| \cdot ||$ and is stable with respect to the norm $|| \cdot ||$, then it is convergent of order (p, q) with respect to the norm $|| \cdot ||$.

Proof Let $u = u(x, t)$ denote the exact solution of the initial-value problem. Then since the difference scheme is accurate of order (p, q) we have

$$u^{n+1} = Qu^n + \Delta t\, T^n,$$

with $||T^n|| = \mathcal{O}((\Delta x)^p) + \mathcal{O}((\Delta t)^q)$. Define

$$W^j = u^j - U^j.$$

Then,

$$W^{n+1} = QW^n + \Delta t\, T^n.$$

Applying this equation repeatedly we obtain

$$\begin{aligned}
W^{n+1} &= QW^n + \Delta t\, T^n \\
&= Q(QW^{n-1} + \Delta t\, T^{n-1}) + \Delta t\, T^n \\
&= Q^2 W^{n-1} + \Delta t\, Q T^{n-1} + \Delta t\, T^n \\
&\quad \cdots \\
&= Q^{n+1} W^0 + \Delta t \sum_{j=0}^{n} Q^j T^{n-j}.
\end{aligned}$$

Since $W^0 = 0$, we have

$$W^{n+1} = \Delta t \sum_{j=0}^{n} Q^j T^{n-j}. \tag{3.9}$$

The fact that the difference scheme is stable implies that for any j (see previous proposition)

$$||Q^j|| \le K e^{\beta t}.$$

Taking the norm of both sides of Eq. (3.9) and then using this result yields

$$||W^{n+1}|| \le \Delta t \sum_{j=0}^{n} ||Q^j|| ||T^{n-j}||$$

$$\le \Delta t K \sum_{j=0}^{n} e^{\beta j \Delta t} ||T^{n-j}||$$

$$\le \Delta t K e^{\beta(n+1)\Delta t} \sum_{j=0}^{n} ||T^{n-j}||$$

$$\le (n+1)\Delta t K e^{\beta(n+1)\Delta t} C^*(t) \left((\Delta x)^p + (\Delta t)^q \right),$$

with $C^*(t) = \sup_{0 \le s \le t} C(s)$, where $C(s)$, $s = (n-j)\Delta t$, is the constant involved in the big $\mathcal{O}$ expression for $||T^{n-j}||$. Thus, since $t = (n+1)\Delta t$, as Δx, $\Delta t \to 0$

$$t K e^{\beta t} C^*(t) \left((\Delta x)^p + (\Delta t)^q \right) \to t K e^{\beta t} C^*(t) \times 0 = 0.$$

This is equivalent to $||U^{n+1} - u^{n+1}|| \to 0$. We have

$$||W^{n+1}|| = K(t) \left((\Delta x)^p + (\Delta t)^q \right) = \mathcal{O}\left((\Delta x)^p \right) + \mathcal{O}\left((\Delta t)^q \right),$$

and the convergence is of order (p, q). $\square$

3.4 Matrix Analysis for Stability

In Sect. 3.2, we have presented the von Neumann stability analysis appropriate to study the stability of finite difference methods that determine solutions of problems defined in the whole real line. If the domain is not the whole real line this tool will give us necessary conditions for stability. These necessary conditions are usually very close to the necessary and sufficient stability conditions. This is the reason for which many times we compute the von Neumann stability conditions even if we are not in the real line. Alternative mathematical tools can be used to study the stability of finite difference methods, such as, the determination of the eigenvalues or the norm of the matrix iteration of the finite difference method. This is the approach we discuss in this section.

3.4.1   *Power Matrices*

To compute the numerical solution of finite difference equations in a computational domain, we solve these equations in a finite number of spatial mesh points even in the cases we have a theoretical infinite domain. We advance in time-rows from the initial line, on which initial values are known, until a finite time T. If no rounding errors were introduced in this numerical process the exact solution of the finite difference equations would be obtained at each node (x_j, t_n). The idea of stability is to control this numerical process, since the finite differences will not be computed exactly. To discuss the stability, additionally to the approach presented previously, we can study the behaviour of the system of the finite difference equations.

Consider the linear system of equations

$$U^{n+1} = A_N U^n, \tag{3.10}$$

where A_N is a matrix $N \times N$ and U^n is a vector of size N. The matrix A_N can be a matrix iteration of a finite difference numerical method. The study of the asymptotic behaviour of $||U^n||$ for some norm $|| \cdot ||$ is important for the convergence of the iterative method when solving the linear system and consequently will be also important for the study of the stability of the finite difference method.

From (3.10) we obtain iteratively that

$$U^{n+1} = A_N^n U^0.$$

This implies

$$||U^{n+1}|| \leq ||A_N^n|| \, ||U^0||, \tag{3.11}$$

where we are considering matrix norms induced by the vector norms. From the previous inequality we arrive at

$$||U^{n+1}|| \leq ||A_N||^n ||U^0||. \tag{3.12}$$

By observing (3.11) and (3.12), it is easy to infer that the stability according to Definition 3.7 will depend on the behaviour of $||A_N^n||$ or $||A_N||^n$. Both behaviours are important. However in many cases $||A_N^n||$ is much smaller than $||A_N||^n$.

The consequences of the definitions of practical stability given previously is that a norm of a matrix A_N compatible with a norm of U^n must satisfy

$$||A_N|| \leq 1. \tag{3.13}$$

If there exists any norm in which $||A_N|| < 1$, then (3.12) tell us that

$$\lim_{n \to \infty} ||U^n|| \leq \lim_{n \to \infty} (||A_N||)^n ||U^0|| = 0.$$

We know that vector norms are equivalent and therefore we can conclude that if this asymptotic behaviour is valid for one norm we have $||U^n|| \to 0$ in any norm. Similarly if $||U^n||$ blows up in some norm, then $||U^n||$ blows up in every equivalent norm.

The condition (3.13) is most appropriate when the solution of the differential equation does not increase as t increases. If we want to take in consideration some other cases, the norm of the matrix can satisfy

$$||A_N|| \leq 1 + \mathcal{O}(\Delta t).$$

3.4.2 Eigenvalue Analysis

There are several important results that relate the power matrices with the eigenvalues of the matrix.

Proposition 3.2 *Consider the linear recurrence system (3.10).*

(a) U^n goes to zero, that is, $A_N^n \to 0$, as $n \to \infty$ if and only if $\rho(A_N) < 1$;
(b) U^n is unbounded when there exists an eigenvalue λ such that $\lambda > 1$;
(c) When $\rho(A_N) = 1$, the solution U^n is bounded or unbounded as $n \to \infty$, according as to whether the eigenvalues λ_j for which $\lambda_j = 1$ are semi-simple or include at least one defective eigenvalue.

Recall that an eigenvalue of multiplicity m greater than unity is said to be semi-simple if it admits m linearly independent eigenvectors; otherwise it is said to be defective.

Proof See [32, page 113] for indications on the proof. $\square$

Eigenvalue analysis is particularly useful for normal matrices. Recall that a real matrix A_N is a normal matrix if $A_N A_N^T = A_N^T A_N$. For instance, symmetric matrices are normal matrices.

For general matrices we have a property that relates the norm of a matrix with the spectral radius of a matrix, that is, $\rho(A) \leq ||A||$, for any norm (see Exercise 3.2 for the proof). However, for normal matrices we have $\rho(A_N) = ||A_N||_2$, where $|| \cdot ||_2$ is the induced euclidean norm (see [32, page 32] and Exercise 3.3 for the proof).

The next results tell us how to compute the eigenvalues of a tridiagonal matrix and of a circulant matrix respectively.

Theorem 3.4 *Let A_N be a matrix $N \times N$ tridiagonal with a_0 in the main diagonal, a_1 in the upper diagonal and a_{-1} in the lower diagonal. Then the eigenvalues are given by*

$$\lambda_k^A = a_0 + 2a_1 \sqrt{\frac{a_{-1}}{a_1}} \cos\left(\frac{k\pi}{N+1}\right), \quad k = 1, \dots, N$$

and the corresponding eigenvectors v_k^A have the j-th component

$$\left(v_k^A\right)_j = \left(\sqrt{\frac{a_{-1}}{a_1}}\right)^j \sin\left(\frac{jk\pi}{N+1}\right).$$

Proof The proof can be seen in [126, page 154–156]. $\square$

The previous formulas holds if a_{-1}/a_1 is negative in which case the eigenvalues are complex. Note that if A_N is real then A_N^T has the same eigenvalues as A_N.

Theorem 3.5 *Let A_N be a circulant matrix $N \times N$, that is,*

$$A = \begin{bmatrix} a_1 & a_2 & \ldots & \ldots & a_{N-1} & a_N \\ a_N & a_1 & \ldots & \ldots & a_{N-2} & a_{N-1} \\ & & & & & \\ a_3 & a_4 & \ldots & \ldots & a_1 & a_2 \\ a_2 & a_3 & \ldots & \ldots & a_N & a_1 \end{bmatrix}.$$

Then A_N has the eigenvalues

$$\lambda_k^A = \sum_{j=1}^{N} a_j e^{i2\pi(j-1)k/N}.$$

Proof The proof can be seen in [38, page 72] and [62, page 186]. $\square$

Another useful result in this context is the Gershgorin theorem.

Theorem 3.6 (Gershgorin Theorem) *Let A_N be a matrix of size $N \times N$ and let D_i be the closed disk in the complex plane centered in a_{ii} with radius $r_i = \sum_{j \neq i} |a_{ij}|$, that is, $D_i = \{z \in \mathbb{C} : |z - a_{ii}| \leq r_i\}$. Then*

(a) All the eigenvalues of A_N lie in the union of the disks D_i, $i = 1, 2, \ldots, N$.
(b) If some of k overlapping disks is disjoint from all the other disks, then exactly k eigenvalues lie in the union of these k disks.

Proof The proof can be seen in [151, page 16]. $\square$

A simple criterion for regulating the error growth of (3.10) is given by $\rho(A_N) \leq 1$. When the matrix is not normal the spectral radius gives no indication of the magnitude of U^{n+1} for finite n. In this case the condition $\rho(A_N) \leq 1$ guarantees eventual decay of the solution, but does not control the intermediate growth of the solution. Then it is easy to understand that the condition $\rho(A_N) \leq 1$ is a necessary condition for stability but not always sufficient.

We finish this section with a classical example that illustrates with the help of the von Neumann analysis that the eigenvalues of the matrix iteration does not always give a sufficient condition for stability.

Suppose we have the problem defined by the partial differential equation

$$\frac{\partial u}{\partial t} = -\frac{\partial u}{\partial x}, \quad x \in (0, 1),$$

with an initial condition $u(x, 0) = u_0(x)$ and a Dirichlet boundary condition $u(0, t)$ given for all t.

Consider the discrete domain defined by $x_j = j\Delta x$, $j = 0, 1, \ldots, N + 1$, where Δx is the space step and $x_0 = 0$, $x_{N+1} = 1$. The discretisation in time is $t_n = n\Delta t$, $n = 0, 1, \ldots, M$. We approximate the time and space derivatives with a first order approximation, obtaining the finite difference method,

$$U_j^{n+1} = U_j^n - v(U_j^n - U_{j-1}^n), \qquad \text{where} \quad v = \frac{\Delta t}{\Delta x}. \tag{3.14}$$

The matricial form of the numerical method is

$$\mathbf{U}^{n+1} = A_N \mathbf{U}^n + \mathbf{v},$$

where $\mathbf{U}^n = [U_1^n, \ldots, U_N^n]^T$,

$$A_N = \begin{bmatrix} 1-v & & & \\ v & 1-v & & \\ \vdots & \ddots & & \vdots \\ & & v & 1-v \end{bmatrix} \quad \text{and} \quad \mathbf{v} = \begin{bmatrix} vU_0^n \\ 0 \\ \vdots \\ \vdots \\ 0 \end{bmatrix}.$$

The eigenvalues are $\lambda_i = 1 - v$, $i = 1, \ldots, N$, and $\rho(A_N) \leq 1$ if and only if $0 \leq v \leq 2$.

On the other hand the von Neumann stability analysis for the interior scheme says that the amplification factor is given by

$$\kappa = 1 - v(1 - e^{-i\xi\Delta x}). \tag{3.15}$$

Therefore, for $\theta = \xi\Delta x$,

$$\kappa(\theta) = 1 - v + v\cos\theta - iv\sin\theta.$$

We have

$$|\kappa(\theta)|^2 = 1 - 2v(1 - v)(1 - \cos\theta).$$

Hence $|\kappa(\theta)|^2 \leq 1$ if and only if $0 \leq v(1 - v)(1 - \cos\theta) \leq 1$, that is, $0 \leq v \leq 1$.

In conclusion the von Neumann analysis gives the stability condition

$$0 \leq \nu \leq 1,$$

and the spectral radius suggests stability for

$$0 \leq \nu \leq 2.$$

By running numerical tests we observe that the scheme is unstable for the values of ν between 1 and 2. The solution is not unbounded but we observe the typical oscillations obtained when we violate the von Neumann stability condition in particular when we increase the value of N.

We want to understand why in the previous example the von Neumann analysis is giving us more reliable information than the eigenvalue analysis. When we study the matrix iteration A_N we are analysing the behaviour of a finite matrix. However, theoretically we are interested also in what happens when $\Delta x \to 0$, that is, when $N \to \infty$ and this behaviour may not be predicted by the analysis of finite matrices. The importance of Gershgorin theorems for the localization of eigenvalues is well established but the results we have presented are also only valid for finite matrices.

Hence, we are led to ask whether the von Neumann analysis is in some way related to the spectral radius of the infinite matrix. To see that the von Neumann condition is related to the spectral radius of the infinite iterative matrix we give a short overview about the spectral radius of infinite Toeplitz matrices. The iterative matrix of the previous example is included in this type of matrices and for many other examples of finite difference methods related to diffusion problems the iterative matrices are Toeplitz matrices.

3.4.3   The Spectral Radius of Infinite Toeplitz Matrices

For linear operators on finite dimensional vector spaces, a matrix operator T is singular if and only if 0 is an eigenvalue of T. However, if T is a linear operator on an infinite dimensional vector space, then T can be singular even though 0 is not an eigenvalue. Consequently, in infinite dimensions the study of eigenvalues of a linear operator T is most often replaced by the study of a larger set known as the spectrum.

Definition 3.10 (Wiener Algebra) Let $\mathbf{T} = \{t \in \mathbf{C} : |t| = 1\}$ be the complex unit circle. The Wiener algebra W is defined as the set of all functions $a : \mathbf{T} \to \mathbf{C}$ with absolutely convergent Fourier series, that is, it is the collection of all functions $a : \mathbf{T} \to \mathbf{C}$ which can be represented in the form

$$a(t) = \sum_{n=-\infty}^{\infty} a_n t^n \; (t \in \mathbf{T}) \quad \text{with} \quad ||a||_W = \sum_{n=-\infty}^{\infty} |a_n| < \infty.$$

The previous condition can be written as

$$a(e^{i\theta}) = \sum_{n=-\infty}^{\infty} a_n e^{in\theta} \ (e^{i\theta} \in \mathbf{T}) \quad \text{with} \quad ||a||_W = \sum_{n=-\infty}^{\infty} |a_n| < \infty.$$

Let $T(a) = (\ldots, a_{-2}, a_{-1}, a_0, a_1, a_2 \ldots)$ given by

$$T(a) = \begin{bmatrix} a_0 & a_1 & \cdots & & \cdots \\ a_{-1} & a_0 & a_1 & & \cdots \\ a_{-2} & a_{-1} & a_0 & a_1 & \cdots & \cdots \\ \vdots & \ddots & & & \vdots & \cdots \\ & & & a_{-1} & a_0 & \cdots \end{bmatrix}.$$

The matrix $T(a)$ is called the infinite Toeplitz matrix generated by a, while a is referred to as the symbol of the matrix $T(a)$.

The spectrum of an operator T is the set

$$\mathrm{sp}(T) = \{\lambda \in \mathbf{C} : T - \lambda I \text{ is not invertible}\}.$$

The next result is in [18, page 12]. See also [147].

Theorem 3.7 *Let the spectral radius of $T(a)$ and $||a||_\infty$ be defined respectively as*

$$\mathrm{rad}\, T(a) = \max\{|\lambda| : \lambda \in \mathrm{sp} T(a)\}, \quad ||a||_\infty := \max_{t \in \mathbf{T}} |a(t)|.$$

If $a \in W$ then $\mathrm{rad}\, T(a) = ||T(a)||_2 = ||a||_\infty$.

The previous theorem is the condition that we know as the von Neumann condition. Let us take the last example of Sect. 3.4.2, to verify that this is the von Neumann condition. We have

$$a(t) = (1 - v) + vt^{-1}.$$

For $z = e^{i\theta}$, if we impose $|a(e^{i\theta})| \le 1$ we obtain the von Neumann condition (3.15). Then $|a(t)| < 1$ for $0 < v \le 1$.

For an historical overview on the contributions of von Neumann to the study of infinite matrices see for instance [15].

3.5 The Energy Method

When boundaries are included we can turn to the called energy method to verify stability. The main idea of the energy method is to derive a norm for the solution vector which one can then show increases by a factor no greater than $1 + \mathcal{O}(\Delta t)$

at each time step. This implies stability in this norm according to the definition previously presented. It is called energy method because in certain cases the physical energy of the system provides such a norm. In general, this methodology gives us sufficient conditions for stability. Thus to some extend this method can be complementary to the von Neumann analysis which tends to give necessary conditions for problems with boundaries.

If we work with the difference between the exact solution of the differential equation and the solution of the finite difference scheme, the energy method can directly give an estimate of the error growth and the rate of convergence of the numerical method.

The main points of this method are better explained with examples and therefore we present some examples for partial differential equations with integer order derivatives. We start with the example we have already used in the two previous sections.

Consider the transport equation

$$\frac{\partial u}{\partial t} + \frac{\partial u}{\partial x} = 0$$

with an initial condition $u(x, 0) = u_0(x)$ and boundary condition

$$u(0, t) = 0, \quad t > 0.$$

Let us define the operators

$$\delta_+ U_j := \frac{U_{j+1} - U_j}{\Delta x}, \quad \delta_- U_j := \frac{U_j - U_{j-1}}{\Delta x},$$

$$\delta^2 U_j = \delta_+ \delta_- U_j = \frac{U_{j+1} - 2U_j + U_{j-1}}{(\Delta x)^2}.$$

Consider the set of discrete points

$$\mathbf{S}_0 = \{U \in \mathbb{R}^{N+1} : U = \{U_j\}, U_0 = 0\}.$$

Define the inner product and the respective norm,

$$(U, V)_N = \Delta x \sum_{j=1}^{N} U_j V_j \quad \text{and} \quad \|U\|_N^2 = (U, U)_N. \tag{3.16}$$

This notation is chosen to highlight the inclusion of the value $U_N V_N$ in the sum.

The transport problem can be approximated by

$$U_j^{n+1} = U_j^n - \Delta t \delta_- U_j^n. \tag{3.17}$$

We have seen that this numerical method is stable if and only if $v \leq 1$, with $v = \Delta t / \Delta x$. We now use the energy method to derive a stability condition.

Proposition 3.3 *If $v \leq 1$ then the numerical method (3.17) is stable, that is,* $||U^{n+1}||_N \leq ||U^n||_N$.

Proof Squaring both sides of the numerical method and summing over mesh points we obtain

$$\sum_{j=1}^{N} (U_j^{n+1})^2 = \sum_{j=1}^{N} (U_j^n - \Delta t \delta_- U_j^n)^2$$

and

$$||U^{n+1}||_N^2 = \Delta x \sum_{j=1}^{N} ((1-v)U_j^n + vU_{j-1}^n)^2.$$

Hence,

$$||U^{n+1}||_N^2 = \Delta x \sum_{j=1}^{N} (1-v)^2 (U_j^n)^2 + 2\Delta x (1-v)v \sum_{j=1}^{N} U_j^n U_{j-1}^n$$

$$+ \Delta x v^2 \sum_{j=1}^{N} (U_{j-1}^n)^2. \tag{3.18}$$

By the Cauchy-Schwarz inequality, we have, for the second term on the right hand side of (3.18),

$$\Delta x \sum_{j=1}^{N} U_j^n U_{j-1}^n \leq \left(\Delta x \sum_{j=1}^{N} (U_j^n)^2 \right)^{1/2} \left(\Delta x \sum_{j=1}^{N} (U_{j-1}^n)^2 \right)^{1/2}.$$

Because $U_0^n = 0$, the third term on the right hand side of (3.18) becomes

$$\Delta x \sum_{j=1}^{N} (U_{j-1}^n)^2 = \Delta x \sum_{j=1}^{N-1} (U_j^n)^2 \leq \Delta x \sum_{j=1}^{N} (U_j^n)^2.$$

Therefore, if $v \leq 1$, then

$$||U^{n+1}||_N^2 \leq (1-v)^2 ||U^n||_N^2 + 2(1-v)v||U^n||_N^2 + v^2 ||U^n||_N^2 \leq ||U^n||_N^2$$

and consequently

$$||U^{n+1}||_N^2 \le ||U^n||_N^2.$$

$\square$

Let us now consider the diffusion equation

$$\frac{\partial u}{\partial t} = D\frac{\partial^2 u}{\partial x^2}$$

with an initial condition $u(x, 0) = u_0(x)$ and boundary conditions

$$u(0, t) = u(1, t) = 0.$$

The diffusion equation can be approximated by the explicit Euler numerical method

$$U_j^{n+1} = U_j^n + D\Delta t \delta^2 U_j^n. \tag{3.19}$$

Consider the set of discrete points

$$\mathbf{S}_0^0 = \{U \in \mathbb{R}^{N+1} : U = \{U_j\},\ U_0 = 0;\ U_N = 0\}.$$

Define the inner product and the respective norm

$$(U, V) = \Delta x \sum_{j=1}^{N-1} U_j V_j \quad \text{and} \quad ||U||^2 = (U, U).$$

Define also

$$(U, V)_0 = \Delta x \sum_{j=0}^{N-1} U_j V_j \quad \text{and} \quad ||U||_0^2 = (U, U)_0.$$

This notation is chosen to highlight the inclusion of the value $U_0 V_0$ in the sum.

Before proving the stability using the energy method we state the following properties.

Lemma 3.2 *For $U, V \in \mathbf{S}_0^0$, we have:*

(a) $(\delta_- U, V) = -(U, \delta_+ V);$
(b) $(\delta^2 U, V) = -(\delta_+ U, \delta_+ V)_0;$
(c) $||\delta_- U|| \le ||\delta_+ U||_0;$
(d) $||\delta_+ U||_0 \le \dfrac{2}{\Delta x}||U||;$
(e) $||\delta^2 U|| \le \dfrac{2}{\Delta x}||\delta_+ U||_0.$

Proof The proof is left as an exercise (see Exercise 3.5). $\square$

The next result presents the stability condition for the numerical method (3.19). We begin by defining

$$\mu = D\frac{\Delta t}{\Delta x^2}.$$

Proposition 3.4 *If $\mu \leq 1/2$ then the numerical method (3.19) is stable, that is,* $||U^{n+1}|| \leq ||U^n||$.

Proof Squaring both sides of the numerical method and summing over mesh points we obtain

$$\sum_{j=1}^{N-1}(U_j^{n+1})^2 = \sum_{j=1}^{N-1}(U_j^n + D\Delta t\delta^2 U_j^n)^2.$$

Therefore

$$||U^{n+1}||^2 = ||U^n||^2 + 2D\Delta t\Delta x\sum_{j=1}^{N-1}U_j^n\delta^2 U_j^n + (D\Delta t)^2\Delta x\sum_{j=1}^{N-1}(\delta^2 U_j^n)^2.$$

By Lemma 3.2 (b), (e) we obtain

$$||U^{n+1}||^2 \leq ||U^n||^2 - 2D\Delta t||\delta_+ U^n||_0^2 + (D\Delta t)^2\frac{4}{\Delta x^2}||\delta_+ U^n||_0^2.$$

It follows that

$$||U^{n+1}||^2 \leq ||U^n||^2 + \left(-2D\Delta t + (D\Delta t)^2\frac{4}{\Delta x^2}\right)||\delta_+ U^n||_0^2.$$

Since $\mu \leq 1/2$ we have

$$\left(-2D\Delta t + (D\Delta t)^2\frac{4}{\Delta x^2}\right) \leq 0.$$

Hence

$$||U^{n+1}||^2 \leq ||U^n||^2.$$

$\square$

We can also use the energy method to prove convergence. In order to do that we define the error $e_j^n = u_j^n - U_j^n$, where u_j^n represents the exact solution at (x_j, t_n) and U_j^n represents its approximation.

Proposition 3.5 *If $\mu \leq 1/2$ then numerical method (3.19) is convergent, that is, the error satisfy*

$$||e^{n+1}|| \leq ||e^n|| + C \max_{0 \leq k \leq n+1} ||T^k||,$$

where C does not depend on the time and space steps.

Proof In this case the error satisfies the equation

$$e_j^{n+1} = e_j^n + D\Delta t \delta^2 e_j^n + \Delta t T_j^n,$$

where T_j^n is the local truncation error.

We have

$$||e^{n+1}|| = ||e^n + D\Delta t \delta^2 e^n + \Delta t T^n|| \leq ||e^n + D\Delta t \delta^2 e^n|| + ||\Delta t T^n||.$$

Then using similar techniques to the ones used to prove the stability, if $\mu \leq 1/2$, we have the inequality

$$||e^n + D\Delta t \delta^2 e^n||^2 \leq ||e^n||^2.$$

Hence,

$$||e^n + D\Delta t \delta^2 e^n|| \leq ||e^n||.$$

Therefore

$$||e^{n+1}|| = ||e^n|| + ||\Delta t T^n||.$$

Consequently

$$||e^{n+1}|| = ||e^0|| + (n+1)\Delta t \sum_{k=0}^{n+1} ||T^k|| \leq ||e^0|| + (n+1)\Delta t \max_{0 \leq k \leq n+1} ||T^k||.$$

$\square$

We move now to a problem that only differs from the previous problem, on the boundary condition considered at $x = 0$.

Consider the diffusion equation

$$\frac{\partial u}{\partial t} = D \frac{\partial^2 u}{\partial x^2}$$

with an initial condition

$$u(x, 0) = u_0(x),$$

a Neumann boundary condition at $x = 0$,

$$\frac{\partial u}{\partial x}(0, t) = 0$$

and a Dirichlet boundary condition at $x = 1$,

$$u(1, t) = 0.$$

The diffusion equation is again approximated by the explicit Euler numerical method, for $j = 1, \ldots, N - 1$,

$$U_j^{n+1} = U_j^n + D\Delta t \delta^2 U_j^n. \tag{3.20}$$

At $j = 0$ we assume the discretization (3.20) is satisfied and to eliminate the ghost point U_{-1}^n we take in consideration the boundary condition that gives us $U_{-1}^n = U_0^n$. This comes from assuming a second order approximation for the first order derivative at $x = 0$. We obtain

$$U_0^{n+1} = U_0^n + \frac{D\Delta t}{(\Delta x)^2}(-U_0^n + U_1^n). \tag{3.21}$$

Before proving the stability using the energy method we state similar properties to the ones stated in Lemma 3.2, but now taking in consideration the fact that we may not have $U_0^n = 0$.

Consider the set of discrete points

$$\mathbf{S}^0 = \{U \in \mathbb{R}^{N+1} : U = \{U_j\}, U_N = 0\}.$$

Lemma 3.3 *For $U, V \in \mathbf{S}^0$, we have:*

(a) $(\delta^2 U, V) = -(\delta_+ U, \delta_+ V)_0 - \delta_+ U_0 V_0;$
(b) $||\delta_- U|| \leq ||\delta_+ U||_0.$

Proof The proof is left as an exercise (see Exercise 3.6). $\qquad\qquad\square$

The next result states the stability condition, for the numerical method (3.20)–(3.21), based in the energy method.

Proposition 3.6 *If $\mu \leq 1/2$ then the numerical method (3.20)-(3.21) is stable, that is, $||U^{n+1}||_0 \leq ||U^n||_0.$*

Proof Squaring both sides of the numerical method and summing over mesh points we obtain

$$\Delta x \sum_{j=0}^{N-1} (U_j^{n+1})^2 = \Delta x \left(U_0^n + \frac{D\Delta t}{(\Delta x)^2}(-U_0^n + U_1^n) \right)^2 + \Delta x \sum_{j=1}^{N-1} (U_j^n + D\Delta t \delta^2 U_j^n)^2.$$

Consequently,

$$\|U^{n+1}\|_0^2 = \Delta x \left(U_0^n + \frac{D\Delta t}{\Delta x} \delta_+ U_0 \right)^2$$

$$+ \Delta x \sum_{j=1}^{N-1} \left[(U_j^n)^2 + 2D\Delta t U_j^n \delta^2 U_j^n + (D\Delta t)^2 (\delta^2 U_j^n)^2 \right].$$

Therefore

$$\|U^{n+1}\|_0^2 = \|U^n\|_0^2 + 2D\Delta t U_0 \delta_+ U_0 + \frac{(D\Delta t)^2}{\Delta x}(\delta_+ U_0)^2$$

$$+ 2D\Delta t \Delta x \sum_{j=1}^{N-1} U_j^n \delta^2 U_j^n + (D\Delta t)^2 \Delta x \sum_{j=1}^{N-1} (\delta^2 U_j^n)^2.$$

By Lemma 3.3 (a), we obtain

$$\|U^{n+1}\|_0^2 = \|U^n\|_0^2 + 2D\Delta t U_0 \delta_+ U_0 + \frac{(D\Delta t)^2}{\Delta x}(\delta_+ U_0)^2$$

$$- 2D\Delta t (\|\delta_+ U\|_0^2 + \delta_+ U_0 V_0) + (D\Delta t)^2 \Delta x \sum_{j=1}^{N-1} (\delta^2 U_j^n)^2.$$

We have

$$\Delta x \sum_{j=1}^{N-1} (\delta^2 U_j^n)^2 = \Delta x \sum_{j=1}^{N-1} \left(\frac{\delta_+ U_j - \delta_- U_j}{\Delta x} \right)^2$$

$$= \frac{1}{\Delta x} \sum_{j=1}^{N-1} (\delta_+ U_j)^2 - \frac{2}{\Delta x} \sum_{j=1}^{N-1} (\delta_+ U_j)(\delta_- U_j) + \frac{1}{\Delta x} \sum_{j=1}^{N-1} (\delta_- U_j)^2$$

$$\leq \frac{1}{(\Delta x)^2} \|\delta_+ U\|^2 + \frac{2}{(\Delta x)^2} \|\delta_+ U\|\|\delta_- U\| + \frac{1}{(\Delta x)^2} \|\delta_- U\|^2,$$

where on the second term on the right hand side we have used the Cauchy-Schwarz inequality. Therefore,

$$||U^{n+1}||_0^2 = ||U^n||_0^2 - 2D\Delta t||\delta_+ U||_0^2 + \frac{(D\Delta t)^2}{\Delta x}(\delta_+ U_0)^2$$

$$+ \frac{(D\Delta t)^2}{(\Delta x)^2}(||\delta_+ U||^2 + 2||\delta_+ U||||\delta_- U|| + ||\delta_- U||^2).$$

Now by Lemma 3.3.(b) and because

$$||\delta_+ U||^2 + \Delta x(\delta_+ U_0)^2 = ||\delta_+ U||_0^2$$

$$||\delta_+ U||^2 \leq ||\delta_+ U||_0^2$$

we have

$$||U^{n+1}||_0^2 \leq ||U^n||_0^2 - 2D\Delta t||\delta_+ U||_0^2 + 4\frac{(D\Delta t)^2}{(\Delta x)^2}||\delta_+ U^n||_0^2.$$

Therefore,

$$||U^{n+1}||_0^2 \leq ||U^{n+1}||_0^2 + \left(-2D\Delta t + 4\frac{(D\Delta t)^2}{(\Delta x)^2}\right)||\delta_+ U^n||_0^2.$$

Since $\mu \leq 1/2$ we have

$$-2D\Delta t + \frac{4(D\Delta t)^2}{(\Delta x)^2} \leq 0$$

and consequently

$$||U^{n+1}||_0^2 \leq ||U^n||_0^2.$$

$\square$

To prove the convergence for the case of having a Neumann boundary condition we can proceed as in the case of Dirichlet boundary conditions and obtain the following result.

Proposition 3.7 *If $\mu \leq 1/2$ then numerical method (3.20)-(3.21) is convergent, that is, the error satisfy*

$$||e^{n+1}||_0 \leq ||e^0||_0 + C \max_{0 \leq k \leq n+1} ||T^k||_0,$$

where C does not depend on the time and space steps.

In Chap. 5, we will show how to use the energy method to prove stability and convergence of finite difference methods to solve the fractional diffusion equations with Dirichlet boundary conditions. But first we need to discuss how to derive finite difference methods for fractional diffusion problems defined in the whole real line and this is what we do in the next chapter.

3.6 Exercises and Solutions

Exercises

3.1 Prove the Discrete Parseval's identity. Let

$$||U||_{l_2} = \left(\Delta x \sum_{j=-\infty}^{\infty} |U_j|^2 \right)^{1/2} \quad \text{and} \quad ||\hat{U}||_{L^2} = \left(\int_{-\pi/\Delta x}^{\pi/\Delta x} |\hat{U}(\xi)|^2 d\xi \right)^{1/2}.$$

If $||U||_{l_2}$ is finite, then also $||\hat{U}||_{L^2}$ is finite and

$$||U||_{l_2} = ||\hat{U}||_{L^2}.$$

3.2 Prove that for a real matrix A, we have $\rho(A) \leq ||A||$, for any norm.

3.3 (a) Prove that a real matrix A is normal if and only if there exists a diagonal matrix D and an unitary matrix U such that $A = UDU^T$. The eigenvectors of A form the rows of the unitary matrix U and the diagonal of D contains the eigenvalues of A.

   (b) Prove that if A is normal then

$$||A||_2 = \rho(A),$$

   where $|| \cdot ||_2$ is the matrix norm induced by the euclidean norm.

3.4 Consider the diffusion equation

$$\frac{\partial u}{\partial t} = D \frac{\partial^2 u}{\partial x^2}$$

with an initial condition $u(x, 0) = u_0(x)$.

   (a) Write the explicit Euler scheme for this equation using the central second order operator to approximate the second order derivative.

   (b) Suppose we consider the problem defined in an open domain, that is, $x \in \mathbb{R}$. Using von Neumann analysis determine the stability conditions of the scheme.

(c) Now assume the problem is defined in $[0, 1]$ with the boundary conditions

$$u(0, t) = u(1, t) = 0.$$

Determine the matricial form of the numerical method.

(d) Calculate the eigenvalues of the tridiagonal matrix and calculate the conditions for which the eigenvalues are less or equal than one. Compare it with the conditions obtained in (b).

(e) Calculate the conditions for which the eigenvalues of the iterative matrix are less or equal than one using the Gershgorin theorem. Compare it with the conditions obtained in (d).

3.5 Prove that for $U, V \in \mathbf{S}_0^0$, we have:

(a) $(\delta_- U, V) = -(U, \delta_+ V)$;
(b) $(\delta^2 U, V) = -(\delta_+ U, \delta_+ V)_0$;
(c) $||\delta_- U|| \leq ||\delta_+ U||_0$;
(d) $||\delta_+ U||_0 \leq \dfrac{2}{\Delta x} ||U||$;
(e) $||\delta^2 U|| \leq \dfrac{2}{\Delta x} ||\delta_+ U||_0$.

3.6 Prove that for $U, V \in \mathbf{S}^0$, we have:

(a) $(\delta^2 U, V) = -(\delta_+ U, \delta_+ V)_0 - \delta_+ U_0 V_0$;
(b) $||\delta_- U|| \leq ||\delta_+ U||_0$.

3.7 Consider the diffusion equation

$$\frac{\partial u}{\partial t} = D \frac{\partial^2 u}{\partial x^2}$$

with an initial condition $u(x, 0) = u_0(x)$ and assume the problem is defined in $[0, 1]$ with the boundary conditions

$$\frac{\partial u}{\partial x}(0, t) = u(1, t) = 0.$$

(a) Justify that a numerical method for this problem can be given by

$$U_j^{n+1} = U_j^n - \mu(U_{j+1}^n - 2U_j^n + U_{j-1}^n), \quad j = 1, \ldots, N - 1,$$

$$U_0^{n+1} = U_0^n - \mu(-U_0^n + U_1^n),$$

$$U_N^n = 0,$$

$$U_j^0 = u_0(x_j), \quad j = 1, \ldots, N - 1.$$

(b) Write the matricial form of the numerical method.

(c) Calculate the conditions for which the eigenvalues of the iterative matrix are less or equal than one.
(d) Compare it with the stability conditions obtained with the energy method.

Solutions

3.1 The proof is similar to the Parseval's identity that has been proved in the end of Chap. 2.

3.2 Assume that λ is an eigenvalue of A. Then, there exists a vector v such that $Av = \lambda v$. Using the properties of a norm we have $||Ax|| = |\lambda|||x||$. The definition of the induced norm of A is

$$||A|| = \sup_x \frac{||Ax||}{||x||}.$$

Therefore,

$$|\lambda| = \frac{||Av||}{||v||} \le \sup_x \frac{||Ax||}{||x||} = ||A||.$$

3.3 (a) See the proof in [32, page 361].

(b) According to (a) we have $U^T A U = D$ and $U A^T U^T = D^T$. Thus $A^T A U = A^T U D = U D^T D$. Hence $A^T A = U(D^T D)U^T$. Then $\rho(A^T A) = \rho(D^T D)$, that is $||A||_2 = \rho(A)$.

3.4 (a) The explicit Euler method consists of approximating the first order derivative in time and second order derivative in space respectively by

$$\frac{\partial u}{\partial t} \approx \frac{U_j^{n+1} - U_j^n}{\Delta t}, \qquad \frac{\partial^2 u}{\partial x^2} \approx \frac{U_{j+1}^n - 2U_j^n + U_{j+1}^n}{(\Delta x)^2},$$

to obtain (3.19).

(b) The amplification factor is given by

$$\kappa(\theta) = 1 + \mu(e^{i\theta} - 2 + e^{-i\theta}).$$

Hence,

$$\kappa(\theta) = 1 + 2\mu(\cos(\theta) - 1).$$

Then show that $\mu \le 1/2$ if and only if $|\kappa(\theta)| \le 1$.

(c) The matricial form of the method is given by $U^{n+1} = AU^n$, where A is a tridiagonal matrix with $1 - 2\mu$ in the diagonal and μ in the upper and lower diagonals.

(d) Compute the eigenvalues using the result to compute the eigenvalues for tridiagonal matrices to obtain

$$\lambda = 1 - 2\mu \left(1 - \cos \left(\frac{k\pi}{N+1} \right) \right), \quad k = 1, \ldots, N.$$

We observe this is similar to (b) and that we obtain the same condition.

(e) If now we use the Gershgorin theorem to calculate where the eigenvalues lie, we obtain for the first line of the matrix $|\lambda - (1 - 2\mu)| \leq \mu$ and for all the others $|\lambda - (1 - 2\mu)| \leq 2\mu$. Then consider $\lambda = a + ib$ and verify if $|\lambda|^2 = a^2 + b^2 \leq 1$ when $\mu \leq 1/2$ or if μ can be larger.

3.5 The proofs are done by direct calculation.

(a) We have

$$(\delta_- U, V) = \Delta x \sum_{j=1}^{N-1} \delta_- U_j V_j = \sum_{j=1}^{N-1} (U_j - U_{j-1}) V_j$$

$$= \sum_{j=1}^{N-1} U_j V_j - \sum_{j=1}^{N-1} U_{j-1} V_j.$$

Since $V_N = 0$, it follows

$$(\delta_- U, V) = \sum_{j=1}^{N-1} U_j V_j - \sum_{j=1}^{N} U_{j-1} V_j = \sum_{j=1}^{N-1} U_j V_j - \sum_{j=0}^{N-1} U_j V_{j+1}.$$

Since $U_0 = 0$ then

$$(\delta_- U, V) = \sum_{j=1}^{N-1} U_j (V_j - V_{j+1}) = -\Delta x \sum_{j=1}^{N-1} U_j \delta_+ V_j = -(U, \delta_+ V).$$

(b) We have

$$(\delta^2 U, V) = \Delta x \sum_{j=1}^{N-1} \delta^2 U_j V_j = \sum_{j=1}^{N-1} (\delta_+ U_j - \delta_- U_j) V_j.$$

Since $V_0 = 0$

$$\sum_{j=1}^{N-1} \delta_+ U_j V_j = \sum_{j=0}^{N-1} \delta_+ U_j V_j.$$

Because $V_N = 0$, then

$$\sum_{j=1}^{N-1} \delta_- U_j V_j = \sum_{j=1}^{N} \delta_- U_j V_j$$

and

$$\sum_{j=1}^{N} (U_j - U_{j-1}) V_j = \sum_{j=0}^{N-1} (U_{j+1} - U_j) V_{j+1}.$$

Therefore,

$$(\delta^2 U, V) = \sum_{j=0}^{N-1} \delta_+ U_j (V_j - V_{j+1}) = -(\delta_+ U, \delta_+ V)_0.$$

(c) We have

$$\|\delta_- U\|^2 = \Delta x \sum_{j=1}^{N-1} (\delta_- U_j)^2 = \frac{1}{\Delta x} \sum_{j=1}^{N-1} (U_j - U_{j-1})^2$$

$$= \frac{1}{\Delta x} \sum_{j=0}^{N-2} (U_{j+1} - U_j)^2.$$

Therefore,

$$\|\delta_- U\|^2 \le \frac{1}{\Delta x} \sum_{j=0}^{N-1} (U_{j+1} - U_j)^2 = \|\delta_+ U\|_0^2.$$

(d) We have

$$\|\delta_+ U\|_0^2 = \sum_{j=0}^{N-1} \Delta x (\delta_+ U_j)^2 = \frac{1}{\Delta x} \sum_{j=0}^{N-1} (U_{j+1}^2 - 2U_j U_{j+1} + U_j^2).$$

Since $-2ab \le a^2 + b^2$, then

$$\|\delta_+ U\|_0^2 \le \frac{1}{\Delta x} \sum_{j=0}^{N-1} (2U_{j+1}^2 + 2U_j^2) = \frac{1}{\Delta x} \left(\sum_{j=1}^{N} 2U_j^2 + \sum_{j=0}^{N-1} 2U_j^2 \right).$$

Because $U_0 = U_N = 0$, we obtain

$$||\delta_+ U||_0^2 \leq \frac{4}{(\Delta x)^2} \sum_{j=1}^{N-1} U_j^2 = \frac{4}{(\Delta x)^2} ||U||^2.$$

(e) We have

$$||\delta^2 U||^2 = \sum_{j=1}^{N-1} \Delta x (\delta^2 U_j)^2 = \Delta x \sum_{j=1}^{N-1} \left(\frac{\delta_+ U_j - \delta_- U_j}{\Delta x} \right)^2$$

$$= \frac{1}{\Delta x} \sum_{j=1}^{N-1} (\delta_+ U_j)^2 - 2 \frac{1}{\Delta x} \sum_{j=1}^{N-1} (\delta_+ U_j)(\delta_- U_j)$$

$$+ \frac{1}{\Delta x} \sum_{j=1}^{N-1} (\delta_- U_j)^2.$$

By the Cauchy-Schwarz inequality and the previous inequality in (c)

$$- \Delta x \sum_{j=1}^{N-1} (\delta_+ U_j)(\delta_- U_j) \leq ||\delta_+ U|| ||\delta_- U|| \leq ||\delta_+ U||_0^2.$$

Consequently

$$||\delta^2 U||^2 \leq \frac{4}{(\Delta x)^2} ||\delta_+ U||_0^2.$$

3.6 (a) We have

$$(\delta^2 U, V) = \Delta x \sum_{j=1}^{N-1} \delta^2 U_j V_j = \sum_{j=1}^{N-1} (\delta_+ U_j - \delta_- U_j) V_j.$$

Since $V_0 \neq 0$

$$\sum_{j=1}^{N-1} \delta_+ U_j V_j = \sum_{j=0}^{N-1} \delta_+ U_j V_j - \delta_+ U_0 V_0.$$

Because $V_N = 0$, then

$$\sum_{j=1}^{N-1} \delta_- U_j V_j = \sum_{j=1}^{N} \frac{(U_j - U_{j-1})}{\Delta x} V_j = \sum_{j=0}^{N-1} \frac{(U_{j+1} - U_j)}{\Delta x} V_{j+1}$$

$$= \sum_{j=0}^{N-1} \delta_+ U_j V_{j+1}.$$

Therefore,

$$(\delta^2 U, V) = \sum_{j=0}^{N-1} \delta_+ U_j (V_j - V_{j+1}) - \delta_+ U_0 V_0 = -(\delta_+ U, \delta_+ V)_0 - \delta_+ U_0 V_0.$$

(b) Similar to Exercise 3.5 (c).

3.7 (a) Explain how we approximate the dervatives in time and space.

(b) The matricial form is $U^{n+1} = AU^n$, where A is a tridiagonal matrix and the first line is different from the others. Include the row $j = 0$ in the matricial form.

(c) Use the result for tridiagonal matrices to compute the eigenvalues.

(d) Compare with the condition $\mu \leq 1/2$.

Chapter 4
Fractional Differential Equations in Unbounded Domains

In this chapter, we present finite difference methods for solving the superdiffusive model introduced in the second chapter, which is related to Lévy flights. The problem we focus on is the Cauchy problem, defined on the real line and subject to an initial condition. We begin by introducing a first order numerical method, and then, to achieve a higher order of accuracy, we introduce another type of approximation, which is second order, for the fractional integrals and derivatives. Other types of approximations can be found in the existing literature. However, our study aims to illustrate the general approach of finite difference methods when applied to fractional diffusion problems.

4.1 Introduction

The fractional advection diffusion equation involves a fractional operator of order $\alpha \in (1, 2)$. Essentially, it is similar to the classical advection diffusion equation, but now this fractional operator replaces the second order derivative. This means that when $\alpha = 2$ in the fractional differential equation, we obtain the classical advection diffusion equation. This type of problems have appeared recently in many physical contexts and applications, that can be find in [10, 11, 14, 19, 30, 34, 56, 91, 111, 125, 176, 177], just to name a few.

The study of the well-posedness of such problems, that is, the existence, uniqueness and regularity of the solutions is still also a very active research field [24, 36, 76]. In the last years, regarding finite difference methods for fractional advection diffusion problems, many approaches have appeared, with some for only pure diffusive problems [29, 106, 108, 145, 167] and others for the advection-diffusion models [27, 71, 84, 123, 129, 131]. Other numerical approaches have been introduced to handle fractional advection-diffusion models [9, 28, 61, 63, 85, 153, 164, 165, 174]. In addition to theoretical investigations, experimental work has been

E. Sousa, *Finite Difference Methods for Fractional Diffusion Equations*, Lecture Notes in Mathematics 2389, https://doi.org/10.1007/978-3-032-11222-4_4

conducted to estimate the parameter α and to characterize the fluid flow behaviour [31, 40, 67, 169].

We derive finite difference methods for a problem defined in the whole real line and discuss the consistency and stability according to the theoretical analysis introduced previously. We have seen that the solutions can be very smooth for the problem defined in the open domain and therefore it is adequate to assume some regularity of the solution when proving the consistency. Since the problem is defined in the whole real line, the von Neumann analysis is a suitable tool to discuss the stability of the numerical methods presented in this chapter.

Let us define the left and right Riemann-Liouville fractional derivatives of order $\alpha \in (1, 2)$. They are respectively defined by

$$\frac{\partial^\alpha u}{\partial x^\alpha}(x, t) = \frac{1}{\Gamma(2 - \alpha)} \frac{\partial^2}{\partial x^2} \int_{-\infty}^{x} u(\tau, t)(x - \tau)^{1-\alpha} d\tau, \tag{4.1}$$

$$\frac{\partial^\alpha u}{\partial(-x)^\alpha}(x, t) = \frac{1}{\Gamma(2 - \alpha)} \frac{\partial^2}{\partial x^2} \int_{x}^{\infty} u(\tau, t)(\tau - x)^{1-\alpha} d\tau. \tag{4.2}$$

The left and right Riemann-Liouville fractional integrals are respectively defined by

$$I_{+}^{2-\alpha} u(x, t) = \frac{1}{\Gamma(2 - \alpha)} \int_{-\infty}^{x} u(\tau, t)(x - \tau)^{1-\alpha} d\tau, \tag{4.3}$$

$$I_{-}^{2-\alpha} u(x, t) = \frac{1}{\Gamma(2 - \alpha)} \int_{x}^{\infty} u(\tau, t)(\tau - x)^{1-\alpha} d\tau. \tag{4.4}$$

We can rewrite the fractional derivatives in terms of the fractional integrals as

$$\frac{\partial^\alpha u}{\partial x^\alpha}(x, t) = \frac{\partial^2}{\partial x^2} I_{+}^{2-\alpha} u(x, t),$$

$$\frac{\partial^\alpha u}{\partial(-x)^\alpha}(x, t) = \frac{\partial^2}{\partial x^2} I_{-}^{2-\alpha} u(x, t).$$

We study finite difference methods to solve the fractional advection diffusion equation, defined for $\alpha \in (1, 2)$ and $\beta \in [-1, 1]$ as

$$\frac{\partial u}{\partial t}(x, t) + V \frac{\partial u}{\partial x}(x, t) = D \nabla_\beta^\alpha u(x, t), \tag{4.5}$$

where

$$\nabla_\beta^\alpha u(x, t) = \frac{1}{2}(1 + \beta)\frac{\partial^\alpha u}{\partial x^\alpha}(x, t) + \frac{1}{2}(1 - \beta)\frac{\partial^\alpha u}{\partial(-x)^\alpha}(x, t). \tag{4.6}$$

The problem is defined in the whole real line and subject to an initial condition. To run the numerical methods for these problems with need to define a computational domain. Therefore, we need to impose computational boundaries. These need to be zero according to the assumption that when $|x| \to \infty$ the solution u goes to zero.

4.2 First Order Finite Difference Methods

We start to present finite difference schemes which are first order in space. They are based on a first order approximation of the fractional derivative, obtained from the Grünwald-Letnikov definition. The approximation we present here have been discussed in several works, such as, [89, 131, 134]. In this chapter, we discuss what happens with this discretisation when the domain is the real line and the case of having a bounded domain will be discussed in the next chapter. Then, we present numerical methods to solve the advection diffusion equation based on this approximation. We present explicit and implicit numerical approximations and we discuss the previous concepts of consistency, stability and convergence.

4.2.1 Grünwald-Letnikov Fractional Derivative Approximation

Different discretizations to approximate the fractional derivatives have been appearing in literature, specially first order discretizations and second order discretizations. We start to present the Grünwald-Letnikov approximation which is very well known in literature. This approximation comes straightforward from the definition of the Grünwald-Letnikov fractional derivative, in the same way that the first order upwind approximation comes from the definition of the standard derivative. More specifically, a straight upwind approximation is when we consider a fixed space step in the definition of the standard derivative instead of considering the limit of the formula when the space step goes to zero. The same can be done to obtain an approximation for the Riemann-Liouville fractional derivative, but using the Grünwald-Letnikov fractional definition.

The Grünwald-Letnikov approximation was one of the first algorithms given to approximate the fractional derivative (see [104]) and it has been proved to be first order accurate under certain conditions. The order of accuracy will depend on the regularity of the function we are considering and on its domain, namely, if the domain is or not the real line.

Before introducing the numerical methods and the subsequent discussion about its truncation error, we need to understand how accurate is an approximation of the fractional derivative. For clarity purposes, we assume we have a function of one variable, $u(x)$, although in the next sections we will be dealing with partial derivatives of a function $u(x, t)$.

We denote by $D_+^\alpha u$ and $D_-^\alpha u$ the left and right Riemann-Liouville fractional derivatives. In a previous chapter we have seen that under certain conditions these derivatives can be defined through the Fourier transform, that is,

$$D_+^\alpha u(x) = \mathscr{F}^{-1}((-i\xi)^\alpha \mathscr{F}(u))(x),$$

$$D_-^\alpha u(x) = \mathscr{F}^{-1}((i\xi)^\alpha \mathscr{F}(u))(x).$$

The next result shows under which conditions the approximation of the fractional derivative by the Grünwald-Letnikov approximation is first order accurate.

Theorem 4.1 *Let* $\alpha \in (1, 2)$ *and consider that the functions* $u, u^{(2)}$ *and* $|\xi|^{\alpha+1}\mathscr{F}(u)$ *belong to* $L_1(\mathbb{R})$. *Let*

$$A_{\Delta x}^+ u(x) = \frac{1}{(\Delta x)^\alpha} \sum_{k=0}^{\infty} (-1)^k \binom{\alpha}{k} u(x - k\Delta x + p\Delta x),$$

and

$$A_{\Delta x}^- u(x) = \frac{1}{(\Delta x)^\alpha} \sum_{k=0}^{\infty} (-1)^k \binom{\alpha}{k} u(x + k\Delta x - p\Delta x),$$

for p *a non-negative integer.*
  If $I_+^{2-\alpha} u \in AC^1(\mathbb{R})$, *then, for a sufficiently small* Δx, *we have, for all* $x \in \mathbb{R}$,

$$A_{\Delta x}^+ u(x) = D_+^\alpha u(x) + \mathscr{O}(\Delta x).$$

Similarly, if $I_-^{2-\alpha} u \in AC^1(\mathbb{R})$, *then, for a sufficiently small* Δx, *we have, for all* $x \in \mathbb{R}$,

$$A_{\Delta x}^- u(x) = D_-^\alpha u(x) + \mathscr{O}(\Delta x).$$

Proof We only present the proof of the first equality, related to the left Riemann-Liouville fractional derivative, since the second equality can be proved in a similar manner. Considering the Fourier transform of the operator $A_{\Delta x}^+ u(x)$, we obtain

$$\mathscr{F}(A_{\Delta x}^+ u)(\xi) = \frac{1}{(\Delta x)^\alpha} \sum_{k=0}^{\infty} (-1)^k \binom{\alpha}{k} e^{i\xi(k-p)\Delta x} \mathscr{F}(u)(\xi),$$

since

$$\mathscr{F}(u(x - \Delta x))(\xi) = e^{i\xi \Delta x} \mathscr{F}(u)(\xi).$$

Now taking in consideration that

$$(1+z)^\alpha = \sum_{k=0}^{\infty} \binom{\alpha}{k} z^k,$$

we obtain

$$\mathscr{F}(A_{\Delta x}^+ u)(\xi) = \frac{1}{(\Delta x)^\alpha}(1 - e^{i\xi \Delta x})^\alpha e^{-i\xi p \Delta x} \mathscr{F}(u)(\xi).$$

We can rewrite it as

$$\mathscr{F}(A_{\Delta x}^+ u)(\xi) = (-i\xi)^\alpha \left(\frac{1 - e^{i\xi \Delta x}}{-i\xi \Delta x}\right)^\alpha e^{-i\xi p \Delta x} \mathscr{F}(u)(\xi).$$

Let

$$w(-i\xi \Delta x) = \left(\frac{1 - e^{i\xi \Delta x}}{-i\xi \Delta x}\right)^\alpha e^{-i\xi p \Delta x}.$$

We can write

$$\mathscr{F}(A_{\Delta x}^+ u)(\xi) = (-i\xi)^\alpha w(-i\xi \Delta x) \mathscr{F}(u)(\xi).$$

Hence,

$$\mathscr{F}(A_{\Delta x}^+ u)(\xi) = (-i\xi)^\alpha \mathscr{F}(u)(\xi) + (-i\xi)^\alpha (w(-i\xi \Delta x) - 1) \mathscr{F}(u)(\xi).$$

We have seen, in a previous chapter, that when $u \in L_1(\mathbb{R})$, $I_+^{2-\alpha} u \in AC^1(\mathbb{R})$ and $u^{(2)} \in L_1(\mathbb{R})$, we have

$$\mathscr{F}(D_+^\alpha u)(\xi) = (-i\xi)^\alpha \mathscr{F}(u)(\xi).$$

Therefore,

$$\mathscr{F}(A_{\Delta x}^+ u)(\xi) = \mathscr{F}(D_+^\alpha u)(\xi) + \mathscr{F}(\phi)(\xi, \Delta x),$$

where

$$\mathscr{F}(\phi)(\xi, \Delta x) = (-i\xi)^\alpha (w(-i\xi \Delta x) - 1) \mathscr{F}(u)(\xi).$$

If we apply the inverse Fourier transform to the previous equation, we can obtain

$$A_{\Delta x}^+ u(x) = D_+^\alpha u(x) + \phi(x, \Delta x).$$

To obtain the first term on the right-hand side of the equation, we need to have the guaranty that the inverse of the Fourier transform of the fractional derivative exists and we have seen that we need to assume $(-i\xi)^{\alpha}\mathscr{F}(u)(\xi) \in L_1(\mathbb{R})$.

Regarding the second term on the right-hand side of the equation, we need to check if the function $\mathscr{F}(\phi)(\xi, \Delta x)$ belongs to $L_1(\mathbb{R})$ in order to guaranty the existence of its inverse Fourier transform. By Taylor expansions we can rewrite, for $z = -i\xi\Delta x$,

$$w(z) = \left(\frac{1 - e^{-z}}{z}\right)^{\alpha} e^{-zp}$$

$$= 1 + zp - \frac{\alpha}{2}z + \mathcal{O}(z^2).$$

Therefore,

$$w(z) - 1 = \left(p - \frac{\alpha}{2}\right)z + \mathcal{O}(z^2).$$

Then, we can conclude that, for a sufficiently small Δx, we have

$$|w(-i\xi\Delta x) - 1| \leq C|\xi|\Delta x,$$

for a positive constant C. Hence,

$$|\mathscr{F}(\phi)(\xi, \Delta x)| = |(-i\xi)^{\alpha}(w(-i\xi\Delta x) - 1)\mathscr{F}(u)(\xi)|$$

$$\leq |(-i\xi)^{\alpha}|C|\xi|\Delta x|\mathscr{F}(u)(\xi)|$$

$$\leq C\Delta x|\xi|^{\alpha+1}|\mathscr{F}(u)(\xi)|.$$

Finally, from this inequality it is easy to conclude that the function $\phi(x, \Delta x)$ is $\mathcal{O}(\Delta x)$, if we have the guaranty that the integral

$$\int_{-\infty}^{\infty} |\xi|^{\alpha+1}|\mathscr{F}(u)(\xi)|d\xi$$

exists. $\square$

Remark 4.1 The hypothesis of the previous theorem can be changed to a different set of functions. If $u \in C^{(4)}(\mathbb{R})$ and $u^{(k)} \in L_1(\mathbb{R})$, $k = 0, 1, \ldots, 4$, the result holds (see Exercise 4.1).

In the next section, we describe how to derive finite difference methods taking in consideration the approximations we have just presented in this section, for the left and right Riemann-Liouville fractional derivatives.

4.2.2   *Implicit Numerical Methods*

Our domain is the real line. To derive a finite difference scheme we suppose there are approximations U_j^n to the values $u(x_j, t_n)$ at the mesh points

$$x_j = j\Delta x, \ \ j = 0, \pm 1, \pm 2, \dots, \quad \text{and} \quad t_n = n\Delta t, \ n \geq 0,$$

where Δx denotes the uniform space step and Δt the uniform time step.

For clarity purposes we start the discussion, on the finite difference methods, for the Eq. (4.5) without advection, that is, $V = 0$. We also consider $\beta = 1$ and therefore we only have the left Riemann-Liouville fractional derivative. The equation with $V = 0$ and $\beta = 1$ becomes

$$\frac{\partial u}{\partial t}(x, t) = D\frac{\partial^\alpha u}{\partial x^\alpha}(x, t). \tag{4.7}$$

We have seen that the left Riemann-Liouville fractional derivative can be approximated by

$$\frac{\partial^\alpha u}{\partial x^\alpha}(x_j, t_n) \approx \sum_{k=0}^{\infty} g_k^\alpha u(x_{j-k}, t_n),$$

where g_k^α are defined as

$$g_k^\alpha = (-1)^k \binom{\alpha}{k}. \tag{4.8}$$

The implicit Euler method, for the diffusion equation (4.7), is

$$\frac{U_j^{n+1} - U_j^n}{\Delta t} = \frac{D}{(\Delta x)^\alpha} \sum_{k=0}^{\infty} g_k^\alpha U_{j-k}^{n+1}.$$

By what we have presented in Sect. 4.2.1 we can conclude easily that this numerical method is consistent (see Exercise 4.2).

Let us analyse the stability based on the von Neumann analysis discussed previously. With that purpose we first give some properties related to the coefficients g_k^α, $k \in \mathbb{N}_0$. The first result is valid for all $\alpha > 0$.

Proposition 4.1 *Let $\alpha > 0$. The coefficients g_k^α, $k \in \mathbb{N}_0$, defined by (4.8), satisfy the following properties.*

(a) The coefficients can be obtained by the recurrence formula

$$g_0^\alpha = 1, \qquad g_{k+1}^\alpha = -\frac{\alpha - k}{k + 1}g_k^\alpha, \ \ k \geq 1. \tag{4.9}$$

(b) $\displaystyle\sum_{k=0}^{\infty} g_k^{\alpha} = 0.$

(c) $\displaystyle\sum_{k=2}^{\infty} g_k^{\alpha} = \alpha - 1.$

Proof

(a) The proof of (a) is left as an exercise at the end of the chapter.
(b) We have that

$$(1 - z)^{\alpha} = \sum_{k=0}^{\infty}(-1)^k \binom{\alpha}{k} z^k.$$

For $z = 1$, we obtain

$$0 = \sum_{k=0}^{\infty}(-1)^k \binom{\alpha}{k}.$$

(c) This result follows directly from (b), since

$$g_0^{\alpha} + g_1^{\alpha} + \sum_{k=2}^{\infty} g_k^{\alpha} = 0,$$

with $g_0^{\alpha} = 1$ and $g_1^{\alpha} = -\alpha$.

$\square$

The next result concerns the properties regarding the sign of the coefficients, for $\alpha \in (1, 2)$.

Proposition 4.2 *Let* $\alpha \in (1, 2)$. *The coefficients* g_k^{α}, $k \in \mathbb{N}_0$, *defined by (4.8), satisfy the following properties.*

(a) $g_0^{\alpha} = 1,\quad g_1^{\alpha} = -\alpha,\quad g_2^{\alpha} = \dfrac{\alpha(\alpha - 1)}{2!}$ *and* $g_k > 0,\quad k \geq 3.$

(b) *For all positive integer N, we have* $\displaystyle\sum_{k=0}^{N} g_k^{\alpha} < 0.$

Proof

(a) After noticing that we can write g_k^{α} as the recurrence formula given in the previous proposition and that we have $g_1^{\alpha} = -\alpha$ and $g_2^{\alpha} = \alpha(\alpha - 1)/2 > 0$, since $\alpha - k < 0$ for $k \geq 3$, by induction we can conclude that $g_k^{\alpha} > 0$ for $k \geq 3$.
(b) Since

$$\sum_{k=0}^{\infty} g_k^{\alpha} = 0$$

and only $g_1^\alpha < 0$, it is easy to conclude that the finite sum starting in 0 until N is negative.

$\square$

The coefficients g_k^α go to zero very quickly and only the first ones are considerably large numbers.

We now proceed to prove that the implicit Euler numerical method is unstable. For the uniform space step Δx and time step Δt, let

$$\mu_\alpha = \frac{D\,\Delta t}{(\Delta x)^\alpha}. \tag{4.10}$$

The quantity μ_α is associated with the diffusion coefficient.

Theorem 4.2 *The implicit Euler numerical method given by*

$$U_j^{n+1} = U_j^n + \mu_\alpha \sum_{k=0}^{\infty} g_k^\alpha U_{j-k}^{n+1}$$

is unstable.

Proof Let

$$U_j^n = (\kappa(\xi))^n e^{i\xi x_j}, \qquad \xi \in \left[-\frac{\pi}{\Delta x}, \frac{\pi}{\Delta x}\right].$$

or

$$U_j^n = (\kappa(\theta))^n e^{ij\theta}, \qquad \theta = \xi\,\Delta x.$$

Then, for the implicit Euler method, it follows

$$(\kappa(\theta))^{n+1} e^{ij\theta} = (\kappa(\theta))^n e^{ij\theta} + \mu_\alpha \sum_{k=0}^{\infty} g_k^\alpha (\kappa(\theta))^{n+1} e^{i(j-k)\theta}.$$

We obtain,

$$\kappa(\theta) = 1 + \mu_\alpha \kappa(\theta) \sum_{k=0}^{\infty} g_k^\alpha e^{-ik\theta}.$$

Therefore,

$$\kappa(\theta) = \frac{1}{1 - \mu_\alpha \sum_{k=0}^{\infty} g_k^\alpha e^{-ik\theta}}.$$

To prove that the method is unstable, it is enough to prove that there exists a θ for which the amplification factor is larger than 1. If we try $\theta = \pi$ we have

$$\sum_{k=0}^{\infty} g_k^{\alpha} e^{-ik\pi} = \sum_{k=0}^{\infty} g_k^{\alpha} \cos(k\theta) = \sum_{k=0}^{\infty} g_k^{\alpha}(-1)^k = 2^{\alpha}.$$

See Exercise 4.3 for the last equality. Then, it is true that $\kappa(\pi) > 1$. $\qquad\qquad\square$

In a similar way to the previous proof, we can prove that the explicit Euler numerical method is also unstable (see Exercise 4.4). Therefore, to obtain stable methods, we need to change the fractional derivative approximation.

Consider a shifted approximation for the fractional derivative,

$$\frac{\partial^{\alpha} u}{\partial x^{\alpha}}(x_j, t_n) \approx \frac{1}{(\Delta x)^{\alpha}} \sum_{k=0}^{\infty} g_k^{\alpha} u(x_{j-k+1}, t_n).$$

We obtain the following result.

Theorem 4.3 *The implicit Euler numerical method given by*

$$U_j^{n+1} = U_j^n + \mu_\alpha \sum_{k=0}^{\infty} g_k^{\alpha} U_{j-k+1}^{n+1}$$

is stable.

Proof Let

$$U_j^n = (\kappa(\theta))^n e^{ij\theta}, \qquad \theta = \xi \Delta x.$$

Then, introducing this node in the implicit Euler method above, we obtain

$$(\kappa(\theta))^{n+1} e^{ij\theta} = (\kappa(\theta))^n e^{ij\theta} + \mu_\alpha \sum_{k=0}^{\infty} g_k^{\alpha} (\kappa(\theta))^{n+1} e^{i(j-k+1)\theta}.$$

Hence,

$$\kappa(\theta) = 1 + \mu_\alpha \kappa(\theta) \sum_{k=0}^{\infty} g_k^{\alpha} e^{-i(k-1)\theta}.$$

Therefore,

$$\kappa(\theta) = \frac{1}{1 - \mu_\alpha \sum_{k=0}^{\infty} g_k^{\alpha} e^{-i(k-1)\theta}}.$$

For $\theta = \pi$, we have

$$\sum_{k=0}^{\infty} g_k^{\alpha} e^{-i(k-1)\pi} = \sum_{k=0}^{\infty} g_k^{\alpha} \cos((k-1)\theta) = \sum_{k=0}^{\infty} g_k^{\alpha}(-1)^{k-1} = -(1+1)^{\alpha} = -2^{\alpha}$$

and consequently

$$|\kappa(\pi)| < 1.$$

This is not anymore an unstable mode.

Let

$$S := \sum_{k=0}^{\infty} g_k^{\alpha} e^{-i(k-1)\theta}.$$

To conclude the numerical method is stable we need to conclude that $|\kappa(\theta)| \leq 1$, for all $\theta \in [-\pi, \pi]$. This will be true, if

$$|1 - \mu_{\alpha} S| \geq 1.$$

that is, if $\mathrm{Re}\,(S) \leq 0$. We have

$$\mathrm{Re}\,(S) = \sum_{k=0}^{\infty} g_k^{\alpha} \cos((k-1)\theta)$$

$$= g_0^{\alpha} \cos(\theta) + g_1^{\alpha} + g_2^{\alpha} \cos(\theta) + \sum_{k=3}^{\infty} g_k^{\alpha} \cos((k-1)\theta).$$

Since $g_k^{\alpha} > 0$, for all $k \neq 0$, and $g_1^{\alpha} < 0$, then

$$\sum_{k=0}^{\infty} g_k^{\alpha} \cos((k-1)\theta) \leq g_0^{\alpha} + g_1^{\alpha} + g_2^{\alpha} + \sum_{k=3}^{\infty} g_k^{\alpha} = 0.$$

$\square$

Consider the Crank-Nicolson method, for the fractional diffusion problem (4.7),

$$\left(1 - \frac{\mu_{\alpha}}{2} \delta_{l,1}^{\alpha}\right) U_j^{n+1} = \left(1 + \frac{\mu_{\alpha}}{2} \delta_{l,1}^{\alpha}\right) U_j^{n},$$

where

$$\delta_{l,1}^{\alpha} U_j^{n} = \sum_{k=0}^{\infty} g_k^{\alpha} U_{j-k+1}^{n}. \tag{4.11}$$

In the notation $\delta_{l,1}^{\alpha}$ of the operator, l stands for the left, because it is related to the approximation of the left Riemann-Liouville fractional derivative and 1 is the order of the approximation.

Since we are in the real line the best way to prove the stability for the Crank-Nicolson method is to use the von Neumann analysis.

Theorem 4.4 *The Crank-Nicolson numerical method given by*

$$\left(1 - \frac{\mu_\alpha}{2}\delta_{l,1}^{\alpha}\right) U_j^{n+1} = \left(1 + \frac{\mu_\alpha}{2}\delta_{l,1}^{\alpha}\right) U_j^n$$

is stable

Proof Let

$$U_j^n = (\kappa(\theta))^n e^{ij\theta}, \qquad \theta = \xi \Delta x.$$

Then, introducing this node in the Crank-Nicolson method above, we obtain

$$(\kappa(\theta))^{n+1} - \frac{\mu_\alpha}{2}\sum_{k=0}^{\infty} g_k^\alpha (\kappa(\theta))^{n+1} e^{i(-k+1)\theta} = (\kappa(\theta))^n + \frac{\mu_\alpha}{2}\sum_{k=0}^{\infty} g_k^\alpha (\kappa(\theta))^n e^{i(-k+1)\theta}.$$

Hence,

$$\kappa(\theta)\left(1 - \frac{\mu_\alpha}{2}\sum_{k=0}^{\infty} g_k^\alpha e^{-i(k-1)\theta}\right) = 1 + \frac{\mu_\alpha}{2}\sum_{k=0}^{\infty} g_k^\alpha e^{-i(k-1)\theta}.$$

Therefore,

$$\kappa(\theta) = \frac{2 + \mu_\alpha \sum_{k=0}^{\infty} g_k^\alpha e^{-i(k-1)\theta}}{2 - \mu_\alpha \sum_{k=0}^{\infty} g_k^\alpha e^{-i(k-1)\theta}}.$$

By observing we have

$$\mathrm{Re}\left(\sum_{k=0}^{\infty} g_k^\alpha e^{-i(k-1)\theta}\right) \leq 0,$$

we can then conclude that

$$|\kappa(\theta)| \leq 1.$$

$\square$

Another way to deal with stability is to study the matricial form of the numerical method. But it is not easy to prove properties for these matrices, which are not symmetric. In literature, we can find many authors that rely on the study of the eigenvalues of the iterative matrices associated with the numerical methods, but as we have seen, this may not be as good as the von Neumann analysis. If the matrices are not normal we can not conclude easily from the eigenvalue condition that the numerical methods are stable.

In the next result we show how the eigenvalue analysis can be applied to study the stability of a numerical method.

Define the computational domain as $[x_0, x_N]$ and assume that the solution of our problem can be considered zero for $x \leq x_0$. The points of the discrete domain are defined by $x_j = x_0 + j\Delta x$, $j = 0, 1, \ldots, N$. Therefore the series (4.11) becomes a finite sum given by

$$\delta_{1,1}^{\alpha,f} U_j^n = \sum_{k=0}^{j+1} g_k^\alpha U_{j-k+1}^n.$$

The notation $\delta_{1,1}^{\alpha,f}$ as now included the f for the finite sum. We consider $U_0^n = 0$, for all n. The matricial form is

$$(I - \frac{\mu_\alpha}{2} A)U^{n+1} = (I + \frac{\mu_\alpha}{2} A)U^n,$$

where $U^n = (U_1^n, \ldots, U_{N-1}^n)^T$ and A is a matrix given by

$$A_{i,j} = \begin{cases} g_{j-k+1}^\alpha & \text{for } k \leq j - 1, \\ g_1^\alpha & \text{for } k \leq j, \\ g_0^\alpha & \text{for } k = j + 1, \\ 0 & \text{for } k > j + 1. \end{cases}$$

We frequently find the following approach in literature, that is, the stability is concluded based in the fact that the eigenvalues are less than one. But we need to be careful, since this is not sufficient to guaranty stability, as illustrated with a classical example in Sect. 3.4.2.

Theorem 4.5 *The Crank-Nicolson numerical method, given by*

$$\left(1 - \frac{\mu_\alpha}{2} \delta_{1,1}^{\alpha,f}\right) U_j^{n+1} = \left(1 + \frac{\mu_\alpha}{2} \delta_{1,1}^{\alpha,f}\right) U_j^n,$$

has an iterative matrix with eigenvalues less than one.

Proof We first show that the eigenvalues of the matrix A have negative real parts. Note that $g_1^\alpha = -\alpha$ and that for $k \geq 1$, we have $g_k^\alpha > 0$. Additionally, for $N > 1$, we have

$$- g_1^\alpha = \alpha \geq g_0^\alpha + \sum_{k=2}^{N} g_k^\alpha.$$

According to the Gershgorin theorem, the eigenvalues of the matrix A are in disks centred at each diagonal entry $A_{j,j} = g_1^\alpha = -\alpha$, with radius

$$r_j = \sum_{k=0}^{N} |A_{j,k}| = \sum_{k=0,\,k\neq j}^{j+1} |g_{j-k+1}^\alpha| < \alpha.$$

These Gershgorin disks are within the left half of the complex plane. Therefore, the eigenvalues of the matrix A have negative real parts.

Next, we observe that λ is an eigenvalue of the matrix A if and only if $(1 - \lambda)$ is an eigenvalue of the matrix $(I - A)$, if and only if $(1 + \lambda)(1 - \lambda)$ is an eigenvalue of the matrix $(I - A)^{-1}(I + A)$. Additionally, the first part of this statement implies that all eigenvalues of the matrix $I - A$ have a magnitude larger than 1 and thus this matrix is invertible. Furthermore, since the real part of λ is negative, it is not hard to check that

$$\left| \frac{1 + \lambda}{1 - \lambda} \right| < 1.$$

Therefore, the spectral radius of the matrix $(I - A)^{-1}(I + A)$ is less than one.   □

Of course for the case $\alpha = 2$ this result guaranties stability because the matrix A is tridiagonal and it is relatively easy to prove that $(I - A)^{-1}(I + A)$ is a normal matrix. But this is not the case for the other values of α.

Consider now the more general problem with $V \neq 0$ and $\beta \in [-1, 1]$. To describe a numerical method for this problem, we need to approximate the first order derivative and we will use the central operator, defined by

$$\Delta_0 U_j^n := \frac{1}{2}(U_{j+1}^n - U_{j-1}^n). \tag{4.12}$$

A discrete approximation to the fractional derivative terms is again defined from the shifted Grünwald-Letnikov formulae,

$$\left(\frac{\partial^\alpha u}{\partial x^\alpha} \right)_j^n \simeq \frac{1}{(\Delta x)^\alpha} \sum_{k=0}^{\infty} g_k^\alpha U_{j+1-k}^n, \tag{4.13}$$

$$\left(\frac{\partial^\alpha u}{\partial (-x)^\alpha} \right)_j^n \simeq \frac{1}{(\Delta x)^\alpha} \sum_{k=0}^{\infty} g_k^\alpha U_{j-1+k}^n. \tag{4.14}$$

We now consider that the fractional operator ∇_β^α, defined by (4.6), is approximated by

$$\frac{\delta_{\beta,1}^\alpha}{2\Delta x^\alpha},$$

where $\delta^\alpha_{\beta,1}$ is given by

$$\delta^\alpha_{\beta,1} U^n_j = (1 + \beta) \sum_{k=0}^\infty g^\alpha_k U^n_{j+1-k} + (1 - \beta) \sum_{k=0}^\infty g^\alpha_k U^n_{j-1+k}. \tag{4.15}$$

The notation $\delta^\alpha_{\beta,1}$ of the operator now includes the parameter β that determines the combination of the left and right Riemann-Liouville fractional derivatives and again 1 stands for the order of accuracy of the approximation.

The Crank-Nicolson method for the general problem is defined by

$$\left(1 + \frac{\nu}{2} \Delta_0 - \frac{\mu_\alpha}{2} \delta^\alpha_{\beta,1}\right) U^{n+1}_j = \left(1 - \frac{\nu}{2} \Delta_0 + \frac{\mu_\alpha}{2} \delta^\alpha_{\beta,1}\right) U^n_j. \tag{4.16}$$

The next result presents the truncation error associated with the numerical method.

Theorem 4.6 *Let* $\alpha \in (1, 2)$ *and* $u(x, \cdot) \in C^3(0, T)$. *For each* t, $u(\cdot, t) \in C^4(\mathbb{R})$, $u(\cdot, t)$, $u_{xx}(\cdot, t)$, $|\xi|^{\alpha+1} \mathscr{F}(u(\cdot, t)) \in L_1(\mathbb{R})$ *and* $I^{2-\alpha}_+ u(\cdot, t)$, $I^{2-\alpha}_- u(\cdot, t) \in AC^1(\mathbb{R})$. *The truncation error of the Crank-Nicolson numerical method (4.16) is of order* $\mathcal{O}(\Delta x) + \mathcal{O}((\Delta t)^2)$.

Proof Let u be a solution of the fractional partial differential equation (4.5) and satisfying the conditions of Theorem 4.1, which are described in the statement of the theorem. Note that the truncation error for the numerical method (4.16) is given by

$$T^n_j = \frac{u^{n+1}_j - u^n_j}{\Delta t} + \frac{V}{\Delta x}\left(\frac{1}{2}\Delta_0 u^{n+1}_j + \frac{1}{2}\Delta_0 u^n_j\right)$$
$$- \frac{D}{(\Delta x)^\alpha}\left(\frac{1}{2}\delta_{\beta,1} u^{n+1}_j + \frac{1}{2}\delta_{\beta,1} u^n_j\right).$$

For $u(x, \cdot) \in C^3(0, T)$ and $u(\cdot, t) \in C^4(\mathbb{R})$, we have

$$\frac{u^{n+1}_j - u^n_j}{\Delta t} = \frac{\partial u(x_j, t_n)}{\partial t} + \frac{\Delta t}{2}\frac{\partial^2 u(x_j, t_n)}{\partial t^2} + \mathcal{O}((\Delta t)^2), \tag{4.17}$$

$$\frac{\Delta_0 u^n_j}{\Delta x} = \frac{\partial^2 u(x_j, t_n)}{\partial x^2} + \mathcal{O}((\Delta x)^2). \tag{4.18}$$

According to Theorem 4.1, if $u(\cdot, t)$, $u_{xx}(\cdot, t)$, $|\xi|^{\alpha+1} \mathscr{F}(u(\cdot, t)) \in L_1(\mathbb{R})$ and $I^{2-\alpha}_+ u(\cdot, t)$, $I^{2-\alpha}_- u(\cdot, t) \in AC^1(\mathbb{R})$, then

$$\frac{\delta_{\beta,1} u^n_j}{2(\Delta x)^\alpha} = \nabla^\alpha_\beta u(x_j, t_n) + \mathcal{O}(\Delta x).$$

Therefore,

$$
T_j^n = \frac{\partial u(x_j, t_n)}{\partial t} + \frac{\Delta t}{2} \frac{\partial^2 u(x_j, t_n)}{\partial t^2} + \mathcal{O}((\Delta t)^2)
$$

$$
+ \frac{V}{2} \left(\frac{\partial^2 u(x_j, t_{n+1})}{\partial x^2} + \frac{\partial^2 u(x_j, t_n)}{\partial x^2} + \mathcal{O}((\Delta x)^2) \right)
$$

$$
- \frac{D}{2} \left(\nabla_\beta^\alpha u(x_j, t_{n+1}) + \nabla_\beta^\alpha u(x_j, t_n) + \mathcal{O}(\Delta x) \right).
$$

Hence, using the fractional differential equation, we obtain

$$
T_j^n = \frac{\partial u(x_j, t_n)}{\partial t} + \frac{\Delta t}{2} \frac{\partial^2 u(x_j, t_n)}{\partial t^2} - \frac{1}{2} \frac{\partial u(x_j, t_n)}{\partial t} - \frac{1}{2} \frac{\partial u(x_j, t_{n+1})}{\partial t}
$$

$$
+ \mathcal{O}((\Delta t)^2) + \mathcal{O}((\Delta x)^2).
$$

Finally, by doing again a Taylor expansion around t_n, of the partial derivative of u in t_{n+1}, we obtain

$$
T_j^n = \mathcal{O}((\Delta t)^2) + \mathcal{O}(\Delta x).
$$

$\square$

The stability of this numerical method can be proved using the von Neumann analysis.

Theorem 4.7 *The Crank-Nicolson numerical method given by*

$$
\left(1 + \frac{v}{2} \Delta_0 - \frac{\mu_\alpha}{2} \delta_{\beta,1}^\alpha \right) U_j^{n+1} = \left(1 - \frac{v}{2} \Delta_0 + \frac{\mu_\alpha}{2} \delta_{\beta,1}^\alpha \right) U_j^n
$$

is stable.

Proof Let

$$
U_j^n = (\kappa(\theta))^n e^{ij\theta}, \qquad \theta = \xi \Delta x.
$$

Introducing the Fourier node in the Crank-Nicolson method above, we obtain

$$
(\kappa(\theta))^{n+1} + (\kappa(\theta))^{n+1} \frac{v}{4} (e^{i\theta} - e^{-i\theta})
$$

$$
- \frac{\mu_\alpha}{2} (\kappa(\theta))^{n+1} \sum_{k=0}^{\infty} g_k^\alpha ((1+\beta) e^{i(-k+1)\theta} + (1-\beta) e^{i(k-1)\theta})
$$

$$
= (\kappa(\theta))^n - (\kappa(\theta))^n \frac{v}{4} (e^{i\theta} - e^{-i\theta})
$$

$$
+ \frac{\mu_\alpha}{2} (\kappa(\theta))^n \sum_{k=0}^{\infty} g_k^\alpha ((1+\beta) e^{i(-k+1)\theta} + (1-\beta) e^{i(k-1)\theta}).
$$

Hence,

$$\kappa(\theta)\left(1 + \frac{\nu}{4}(e^{i\theta} - e^{-i\theta}) - \frac{\mu_\alpha}{2}\sum_{k=0}^{\infty} g_k^\alpha((1+\beta)e^{i(-k+1)\theta} + (1-\beta)e^{i(k-1)\theta})\right)$$

$$= 1 - \frac{\nu}{4}(e^{i\theta} - e^{-i\theta}) + \frac{\mu_\alpha}{2}\sum_{k=0}^{\infty} g_k^\alpha((1+\beta)e^{i(-k+1)\theta} + (1-\beta)e^{i(k-1)\theta}).$$

Therefore,

$$\kappa(\theta) = \frac{2 + i\nu\sin(\theta) + \mu_\alpha\sum_{k=0}^{\infty} g_k^\alpha((1+\beta)e^{i(-k+1)\theta} + (1-\beta)e^{i(k-1)\theta})}{2 - i\nu\sin(\theta) - \mu_\alpha\sum_{k=0}^{\infty} g_k^\alpha((1+\beta)e^{i(-k+1)\theta} + (1-\beta)e^{i(k-1)\theta})}.$$

Observing that

$$(1+\beta)e^{i(-k+1)\theta} + (1-\beta)e^{i(k-1)\theta} = 2\cos((-k+1)\theta) + i2\beta\sin((-k+1)\theta),$$

we can write the amplification factor $\kappa(\theta)$ as

$$\frac{2 + 2\mu_\alpha\sum_{k=0}^{\infty} g_k^\alpha\cos((-k+1)\theta) + i\left[\nu\sin(\theta) + 2\mu_\alpha\beta\sum_{k=0}^{\infty} g_k^\alpha\sin((-k+1)\theta)\right]}{2 - 2\mu_\alpha\sum_{k=0}^{\infty} g_k^\alpha\cos((-k+1)\theta) - i\left[\nu\sin(\theta) + 2\mu_\alpha\beta\sum_{k=0}^{\infty} g_k^\alpha\sin((-k+1)\theta)\right]}.$$

The imaginary parts of the numerator and denominator are equal in modulus and we have seen previously that

$$\sum_{k=0}^{\infty} g_k^\alpha\cos((-k+1)\theta) \leq 0.$$

Therefore, the modulus of the denominator is larger than the modulus of the numerator. We can then conclude that $|\kappa(\theta)| \leq 1$. $\qquad\square$

4.2.3 Explicit Numerical Methods

Remember that our domain is the real line. To derive a finite difference scheme we suppose there are approximations U_j^n to the values $u(x_j, t_n)$ at the mesh points

$$x_j = j\Delta x, \ j \in \mathbb{Z} \quad \text{and} \quad t_n = n\Delta t, \ n \geq 0,$$

where Δx denotes the uniform space step and Δt the uniform time step. For the uniform space step Δx and time step Δt, let again consider the quantities

$$\nu = \frac{V\Delta t}{\Delta x} \quad \text{and} \quad \mu_\alpha = \frac{D\Delta t}{(\Delta x)^\alpha}. \tag{4.19}$$

To describe the finite difference schemes, we use the upwind, central and second difference operators, defined respectively by

$$\Delta_- U_j^n := U_j^n - U_{j-1}^n, \tag{4.20}$$

$$\Delta_0 U_j^n := \frac{1}{2}(U_{j+1}^n - U_{j-1}^n), \tag{4.21}$$

$$\delta^2 U_j^n := U_{j+1}^n - 2U_j^n + U_{j-1}^n. \tag{4.22}$$

Discrete approximations to the fractional derivatives are defined from the shifted Grünwald-Letnikov formulae as previously

$$\left(\frac{\partial^\alpha u}{\partial x^\alpha}\right)_j^n \simeq \frac{1}{(\Delta x)^\alpha} \sum_{k=0}^{\infty} g_k^\alpha U_{j+1-k}^n, \tag{4.23}$$

$$\left(\frac{\partial^\alpha u}{\partial(-x)^\alpha}\right)_j^n \simeq \frac{1}{(\Delta x)^\alpha} \sum_{k=0}^{\infty} g_k^\alpha U_{j-1+k}^n. \tag{4.24}$$

The fractional operator ∇_β^α, defined by (4.6), is therefore approximated as

$$\left(\nabla_\beta^\alpha u\right)_j^n \simeq \frac{\delta_{\beta,1}^\alpha}{2(\Delta x)^\alpha},$$

where $\delta_{\beta,1}^\alpha$ is given by

$$\delta_{\beta,1}^\alpha U_j^n = (1+\beta) \sum_{k=0}^{\infty} g_k^\alpha U_{j+1-k}^n + (1-\beta) \sum_{k=0}^{\infty} g_k^\alpha U_{j-1+k}^n. \tag{4.25}$$

An explicit finite difference scheme to approximate (4.5) can therefore be of the form

$$U_j^{n+1} = U_j^n - \nu\Delta_- U_j^n + \frac{1}{2}\mu_\alpha \delta_{\beta,1}^\alpha U_j^n. \tag{4.26}$$

Here, an upwind discretisation for the advective term is considered and henceforth we call this scheme the *upwind scheme*.

Another approximation can be obtained by replacing the upwind operator with the central operator, such as,

$$U_j^{n+1} = U_j^n - \nu\Delta_0 U_j^n + \frac{1}{2}\mu_\alpha \delta_{\beta,1}^\alpha U_j^n. \tag{4.27}$$

We call this scheme the *central scheme*, according to the discretisation of the advective term.

The third scheme is derived in a similar way to the Lax-Wendroff scheme [80] and therefore we call this scheme *Lax-Wendroff type scheme*. We explain how we obtain this method in what follows, as presented in [129].

Let us expand u about time level n, that is, $t = n\Delta t$, to obtain

$$u(x, t_{n+1}) - u(x, t_n) = \Delta t \frac{\partial u}{\partial t}(x, t_n) + \frac{(\Delta t)^2}{2} \frac{\partial^2 u}{\partial t^2}(x, t_n) + \mathcal{O}((\Delta t)^3). \tag{4.28}$$

Then, from (4.5),

$$\frac{\partial^2 u}{\partial t^2}(x, t) = -V \frac{\partial^2 u}{\partial x \partial t}(x, t) + D\nabla_\beta^\alpha \left(\frac{\partial u}{\partial t}(x, t) \right), \tag{4.29}$$

$$\frac{\partial^2 u}{\partial x \partial t}(x, t) = -V \frac{\partial^2 u}{\partial x^2}(x, t) + D\frac{\partial}{\partial x}\left[\nabla_\beta^\alpha u(x, t) \right]. \tag{4.30}$$

Therefore, a spatial finite-difference approximation can be obtained by dropping out the $\alpha + 1$ and higher-spatial-derivative terms, from the previous equalities, holding

$$\frac{\partial^2 u}{\partial t^2}(x, t) \simeq V^2 \frac{\partial^2 u}{\partial x^2}(x, t). \tag{4.31}$$

Inserting (4.5) and (4.31) into (4.28) gives

$$u(x, t_{n+1}) \simeq u(x, t_n) + \Delta t \left(-V \frac{\partial u}{\partial x}(x, t_n) + D\nabla_\beta^\alpha u(x, t_n) \right)$$
$$+ \frac{1}{2}(\Delta t)^2 \left(V^2 \frac{\partial^2 u}{\partial x^2}(x, t_n) \right). \tag{4.32}$$

Therefore (4.32) can be written in the form

$$u(x, t_{n+1}) \simeq u(x, t_n) - V\Delta t \frac{\partial u}{\partial x}(x, t_n) + \Delta t D\nabla_\beta^\alpha u(x, t_n)$$
$$+ \frac{1}{2}(\Delta t)^2 V^2 \frac{\partial^2 u}{\partial x^2}(x, t_n). \tag{4.33}$$

The difference $(u^{n+1} - u^n)$ becomes

$$u(x, t_{n+1}) - u(x, t_n) = U_j^{n+1} - U_j^n. \tag{4.34}$$

If we discretize the first derivative with the central difference operator, the second order derivative with second order difference operator and the fractional derivative with the respective fractional difference operator, from (4.33), we obtain

$$U_j^{n+1} = U_j^n - \nu\Delta_0 U_j^n + \mu_\alpha \delta_{\beta,1}^\alpha U_j^n + \frac{1}{2}\nu^2 \delta^2 U_j^n. \tag{4.35}$$

We shall deal with functions vanishing at infinity. Therefore, we can choose the computational domain $[x_{-N}, x_N]$, such that, $u(x_{-N}, t)$ and $u(x_N, t)$ are very small. The function u goes to zero, as $|x|$ goes to infinity, and therefore, it is possible to choose an interval in such a way that the truncation of the domain will not interfere with the accuracy expected for the approximated solution.

All the previous methods are explicit and can be written in the form of a matricial system. Assume the nodal points are U_j^n, $j = -N, \ldots, -1, 0, 1, \ldots, N$ and that the boundary conditions are given, which means we know the function values at U_{-N}^n and U_N^n, for $n = 0, 1, 2, \ldots$. Introducing the vector

$$\mathbf{U}^n = [U_{-N+1}^n, \ldots, U_{-1}^n, 0, U_1^n, \ldots, U_{N-1}^n]^T,$$

the schemes may be written as matrix equations

$$\mathbf{U}^{n+1} = M\mathbf{U}^n + \mathbf{v}^n, \; n = 0, 1, 2, \ldots, \tag{4.36}$$

where M is the $(2N - 1) \times (2N - 1)$ matrix iteration and $\mathbf{v}^n$ contains boundary values. In what follows, we write the matrix M and the vector $\mathbf{v}^n$, for the explicit numerical methods. The matrix iteration M has the form

$$M = A + \frac{1}{2}\mu_\alpha B, \tag{4.37}$$

where A and B are matrices of dimension $(2N - 1) \times (2N - 1)$ and A is related with the advection discretisations and B with the diffusion discretisations.

For all the three schemes the matrix B is the same, given by

$$B = (1 + \beta)L + (1 - \beta)L^T,$$

where

$$L = \begin{bmatrix} g_1^\alpha & g_0^\alpha & 0 & \cdots & & 0 \\ g_2^\alpha & & & & & \\ & & & \ddots & & \vdots \\ \vdots & & & \ddots & & 0 \\ & & & & & g_0^\alpha \\ g_{2N-1}^\alpha & & \cdots & & g_2^\alpha & g_1^\alpha \end{bmatrix}.$$

The vector $\mathbf{v}^n$ is composed of two parts $\mathbf{v}^n = \mathbf{v}_A^n + \mathbf{v}_B^n$, where the vector $\mathbf{v}_A^n$ contains the boundary values related to the matrix A and the vector $\mathbf{v}_B^n$ contains the boundary values related to the matrix B.

Now let us have a close look at matrix B. The matricial form show us in a very simple form a theoretical property of the fractional operators that we have not mentioned. The relation between the left fractional operator and the right fractional operator. We see that matricially one is the transpose of the other one. Actually they are adjoint operators. Note also the matrix B is a dense matrix. But we have seen that we are lucky enough to have g_k^α with nice properties. In particular, the coefficients g_k^α go to zero very quickly, only the first ones are considerably large numbers. Therefore, the matrices are not full matrices from the computational point of view, since many of the entries are very close to zero and can safely be assumed as zero.

To analyse the convergence of the numerical methods we study the consistency and the stability. The consistency is studied through the truncation error analysis and we show the schemes are consistent with Eq. (4.5). The study of the stability relies on the von Neumann analysis since we are in the real line.

Let u be the solution of our equation. The fractional operator, ∇_β^α, in all the three schemes is approximated by $\delta_{\beta,1}^\alpha / 2(\Delta x)^\alpha$. From the result presented in Theorem 4.1, taken from [97], under certain conditions, we have that

$$\frac{\delta_{\beta,1}^\alpha u_j^n}{2\Delta x^\alpha} = (\nabla_\beta^\alpha u)_j^n + \mathcal{O}(\Delta x).$$

For the upwind scheme the truncation error is

$$T_j^n = \frac{u_j^{n+1} - u_j^n}{\Delta t} + V \frac{u_j^n - u_{j-1}^n}{\Delta t} - \frac{D}{2(\Delta x)^\alpha} \delta_\beta^\alpha u_j^n$$

$$= \left(\frac{\partial u}{\partial t}\right)_j^n + \mathcal{O}(\Delta t) + V \left(\frac{\partial u}{\partial x}\right)_j^n + \mathcal{O}(\Delta x) - D(\nabla_\beta^\alpha u)_j^n + \mathcal{O}(\Delta x).$$

Therefore, the upwind scheme has an order of accuracy $\mathcal{O}(\Delta t) + \mathcal{O}(\Delta x) + \mathcal{O}(\Delta x)$. Regarding the upwind approximations in space and time of the first order partial derivatives, it is well know that we need to have the respective partial derivatives continuous until the second order.

Similarly, for the central scheme we have

$$T_j^n = \frac{u_j^{n+1} - u_j^n}{\Delta t} + V \frac{u_{j+1}^n - u_{j-1}^n}{\Delta t} - \frac{D}{2(\Delta x)^\alpha} \delta_\beta^\alpha u_j^n$$

$$= \left(\frac{\partial u}{\partial t}\right)_j^n + \mathcal{O}(\Delta t) + V \left(\frac{\partial u}{\partial x}\right)_j^n + \mathcal{O}((\Delta x)^2) - D(\nabla_\beta^\alpha u)_j^n + \mathcal{O}(\Delta x).$$

The order of accuracy is $\mathcal{O}(\Delta t) + \mathcal{O}((\Delta x)^2) + \mathcal{O}(\Delta x)$. For the central approximation to be second order we need to assume the spatial partial derivatives are continuous until the fourth order.

For the Lax-Wendroff scheme, we have

$$T_j^n = \frac{u_j^{n+1} - u_j^n}{\Delta t} + V\frac{u_{j+1}^n - u_{j-1}^n}{\Delta t} + V^2\frac{\Delta t}{2}\frac{u_{j+1}^n - 2u_j^n + u_{j-1}^n}{(\Delta x)^2} - \frac{D}{2(\Delta x)^\alpha}\delta_\beta^\alpha u_j^n.$$

Hence

$$T_j^n = \left(\frac{\partial u}{\partial t}\right)_j^n + \frac{\Delta t}{2}\left(\frac{\partial^2 u}{\partial t^2}\right)_j^n + \mathcal{O}((\Delta t)^2) + V\left(\frac{\partial u}{\partial x}\right)_j^n + \mathcal{O}((\Delta x)^2)$$

$$+ V^2\frac{\Delta t}{2}\left(\frac{\partial^2 u}{\partial x^2}\right)_j^n + \mathcal{O}((\Delta x)^2) - D(\nabla_\beta^\alpha u)_j^n + \mathcal{O}(\Delta x).$$

The order of accuracy is $\mathcal{O}(\Delta t) + \mathcal{O}((\Delta x)^2) + \mathcal{O}(\Delta x)$. For small D, it follows

$$T_j^n = \left(\frac{\partial u}{\partial t}\right)_j^n + \mathcal{O}((\Delta t)^2) + V\left(\frac{\partial u}{\partial x}\right)_j^n + \mathcal{O}((\Delta x)^2) + \mathcal{O}((\Delta x)^2)$$

$$- D(\nabla_\beta^\alpha u)_j^n + \mathcal{O}(\Delta x).$$

The Lax-Wendroff scheme has an order of accuracy close to $\mathcal{O}((\Delta t)^2) + \mathcal{O}((\Delta x)^2) + \mathcal{O}(\Delta x)$. This numerical method achieves this order of accuracy if we assume the spatial partial derivatives are continuous until the fourth order and the time partial derivatives are continuous until the third order. Additionally, we observe that if we have a second order approximation for the fractional derivative, we can obtain a numerical method which is second order accurate. Therefore, this will be discussed in the next section.

In order to derive stability conditions for the finite difference schemes, we apply the von Neumann analysis or Fourier analysis. Recall that although von Neumann analysis gives necessary and sufficient stability conditions only when applied to pure initial value problems and to problems with periodic boundary conditions, it is known that the von Neumann approach always yields necessary conditions for stability even when we have problems with non-periodic boundary conditions.

The von Neumann analysis assumes that any finite mesh function, such as, the numerical solution U_j^n will be decomposed into a Fourier series with general term given by $\kappa^n e^{i\xi(j\Delta x)}$, where κ^n is the amplitude of the harmonic, the product $\xi\Delta x$ is the phase angle θ and covers the domain $[-\pi, \pi]$. Considering a single mode $\kappa^n e^{ij\theta}$, its time evolution is determined by the same numerical scheme as the complete numerical solution U_j^n. Hence inserting a representation of this form into a numerical scheme we obtain stability conditions. The stability conditions will be satisfied if the amplitude factor κ does not grow in time, that is, if we have $|\kappa(\theta)| \leq 1$, for all θ.

Consider the discretisation of the fractional diffusion equation (4.5), when $V = 0$, given by

$$U_j^{n+1} = U_j^n + \frac{1}{2}\mu_\alpha\left[(1+\beta)\sum_{k=0}^{\infty} g_k^\alpha U_{j+1-k} + (1-\beta)\sum_{k=0}^{\infty} g_k^\alpha U_{j-1+k}\right]. \qquad (4.38)$$

For $\alpha = 2$, we have $g_0^\alpha = 1$, $g_1^\alpha = -2$, $g_2^\alpha = 1$ and $g_k^\alpha = 0$ for all $k \geq 3$. Therefore, we obtain the second order discrete operator for the second order derivative.

The following theorem concerns the stability of the numerical scheme (4.38).

Theorem 4.8 *Let $\beta \in [-1, 1]$ and $\alpha \in (1, 2]$. If the numerical scheme (4.38) is von Neumann stable, then $\mu_\alpha \leq 2^{1-\alpha}$.*

Proof If we insert the mode $\kappa^n e^{ij\theta}$ into the scheme (4.38), we obtain the following amplification factor

$$\kappa(\theta) = 1 + \frac{1}{2}\mu_\alpha\left\{(1+\beta)\sum_{k=0}^{\infty} g_k^\alpha e^{i(1-k)\theta} + (1-\beta)\sum_{k=0}^{\infty} g_k^\alpha e^{-i(1-k)\theta}\right\}.$$

Let us consider $\theta = 0$ and $\theta = \pi$. The region around $\theta = 0$ corresponds to the low frequencies, while the region around $\theta = \pi$ is associated with the high-frequencies. In particular, the value $\theta = \pi$ corresponds to the highest frequency resolvable on the mesh, namely frequency of wavelength $2\Delta x$.

For $\theta = 0$ we have

$$\kappa(0) = 1 + \frac{1}{2}\mu_\alpha\left\{(1+\beta)\sum_{k=0}^{\infty} g_k^\alpha + (1-\beta)\sum_{k=0}^{\infty} g_k^\alpha\right\}$$

$$= 1 + \mu_\alpha(g_0^\alpha + g_1^\alpha) + \frac{1}{2}\mu_\alpha\left\{(1+\beta)\sum_{k=2}^{\infty} g_k^\alpha + (1-\beta)\sum_{k=2}^{\infty} g_k^\alpha\right\}$$

$$\leq 1 + \mu_\alpha(1-\alpha) + \frac{1}{2}\mu_\alpha[(1+\beta)(\alpha-1) + (1-\beta)(\alpha-1)] = 1.$$

For $\theta = \pi$, the amplification factor is given by

$$\kappa(\pi) = 1 + \frac{1}{2}\mu_\alpha\left\{(1+\beta)\sum_{k=0}^{\infty} g_k^\alpha \cos((1-k)\pi) + (1-\beta)\sum_{k=0}^{\infty} g_k^\alpha \cos((1-k)\pi)\right\}.$$

Since $\cos((1-k)\pi) = (-1)^{k-1}$, it follows

$$\kappa(\pi) = 1 + \frac{1}{2}\mu_\alpha\left\{-(1+\beta)\sum_{k=0}^{\infty} a_k^\alpha - (1-\beta)\sum_{k=0}^{\infty} a_k^\alpha\right\}$$

$$= 1 - \frac{1}{2}\mu_\alpha\left\{(1+\beta)\sum_{k=0}^{\infty} a_k^\alpha + (1-\beta)\sum_{k=0}^{\infty} a_k^\alpha\right\},$$

where

$$a_k^\alpha = \binom{\alpha}{k}.$$

The condition $|\kappa(\pi)| \le 1$ is equivalent to

$$\frac{1}{2}\mu_\alpha \left\{ (1+\beta) \sum_{k=0}^{\infty} a_k^\alpha + (1-\beta) \sum_{k=0}^{\infty} a_k^\alpha \right\} \le 2.$$

Therefore, since $\sum_{k=0}^{\infty} a_k = 2^\alpha$, we get

$$\frac{1}{2}\mu_\alpha \left\{ (1+\beta) \sum_{k=0}^{\infty} a_k^\alpha + (1-\beta) \sum_{k=0}^{\infty} a_k^\alpha \right\} \le \frac{1}{2}\mu_\alpha[(1+\beta)2^\alpha + (1-\beta)2^\alpha]$$

$$= \mu_\alpha 2^\alpha.$$

Hence, we must have $\mu_\alpha 2^\alpha \le 2$, that is, $\mu_\alpha \le 2^{1-\alpha}$. $\qquad\square$

The following stability results are for the upwind and Lax-Wendroff schemes. The proofs are technically similar to the proof of the previous result. They can be found in [129].

Theorem 4.9 *Let $\beta \in [-1, 1]$ and $\alpha \in (1, 2]$. If the upwind scheme (4.26) is von Neumann stable, then $v + 2^{\alpha-1}\mu_\alpha \le 1$.*

Proof For the upwind scheme the amplification factor is given by

$$\kappa_U(\theta) = 1 - v(1 - e^{-i\theta}) + \frac{1}{2}\mu_\alpha \left\{ (1+\beta) \sum_{k=0}^{\infty} g_k^\alpha e^{i(1-k)\theta} \right.$$

$$\left. + (1-\beta) \sum_{k=0}^{\infty} g_k^\alpha e^{-i(1-k)\theta} \right\}.$$

For $\theta = \pi$, it follows

$$\kappa_U(\pi) = 1 - 2v - \frac{1}{2}\mu_\alpha \left\{ (1+\beta) \sum_{k=0}^{\infty} a_k^\alpha + (1-\beta) \sum_{k=0}^{\infty} a_k^\alpha \right\},$$

with

$$a_k^\alpha = \binom{\alpha}{k}.$$

Therefore, to have $|\kappa_U(\pi)| \leq 1$, we must have

$$2\nu + \mu_\alpha \sum_{k=0}^{\infty} a_k^\alpha \leq 2.$$

Moreover,

$$2\nu + \mu_\alpha \sum_{k=0}^{\infty} a_k^\alpha = 2\nu + \mu_\alpha 2^\alpha$$

and $2\nu + \mu_\alpha 2^\alpha \leq 2$ implies $\nu + \mu_\alpha 2^{\alpha-1} \leq 1$. $\square$

Theorem 4.10 *Let $\beta \in [-1, 1]$ and $\alpha \in (1, 2]$. If the Lax-Wendroff scheme (4.35) is von Neumann stable, then $\nu^2 + 2^{\alpha-1}\mu_\alpha \leq 1$.*

Proof The amplification factor for the Lax-Wendroff scheme is of the form

$$\kappa_{LW}(\theta) = 1 - \frac{\nu}{2}(e^{i\theta} - e^{-i\theta}) + \frac{\nu^2}{2}(e^{i\theta} - 2 + e^{-i\theta})$$

$$+ \frac{1}{2}\mu_\alpha \left\{ (1 + \beta) \sum_{k=0}^{\infty} g_k^\alpha e^{i(1-k)\theta} + (1 - \beta) \sum_{k=0}^{\infty} g_k^\alpha e^{-i(1-k)\theta} \right\}.$$

For $\theta = \pi$, we have

$$\kappa_{LW}(\pi) = 1 - 2\nu^2 - \frac{1}{2}\mu_\alpha \left\{ (1 + \beta) \sum_{k=0}^{\infty} a_k^\alpha + (1 - \beta) \sum_{k=0}^{\infty} a_k^\alpha \right\},$$

with

$$a_k^\alpha = \binom{\alpha}{k}.$$

Similarly, to have $|\kappa_{LW}(\pi)| \leq 1$, we must have

$$2\nu^2 + \mu_\alpha \sum_{k=0}^{\infty} a_k^\alpha \leq 2.$$

Moreover,

$$2\nu^2 + \mu_\alpha \sum_{k=0}^{\infty} a_k^\alpha = 2\nu^2 + \mu_\alpha 2^\alpha$$

and $2\nu^2 + \mu_\alpha 2^\alpha \leq 2$ implies $\nu^2 + \mu_\alpha 2^{\alpha-1} \leq 1$. $\square$

The von Neumann stability conditions are not always easy to obtain as illustrated by these two last numerical methods. We have only obtained necessary conditions. Sufficient and necessary conditions are difficult to obtain.

The derivation of stability conditions for the central scheme in the classical case is usually harder than for the previous schemes. For the central scheme of the standard advection diffusion equation, when $\alpha = 2$, it was also not straightforward. In fact, the stability analysis for the central scheme was controversial for some years because of the apparent difficulty of obtaining necessary and sufficient stability conditions, see for instance [127]. The following theorem concerning the central scheme sets out necessary stability conditions. These conditions are not as strong as the necessary conditions of the previous theorems, in the sense that they are less close to the respective necessary and sufficient stability conditions, as discussed in [129].

Theorem 4.11 *Let $\beta \in [-1, 1]$ and $\alpha \in (1, 2]$. If the central scheme (4.27) is von Neumann stable, then $2^{\alpha-1}\mu_\alpha \leq 1$.*

Proof For the central scheme the amplification factor is given by

$$\kappa_C(\theta) = 1 - \frac{v}{2}(e^{i\theta} - e^{-i\theta})$$

$$+ \frac{1}{2}\mu_\alpha \left\{ (1 + \beta) \sum_{k=0}^{\infty} g_k^\alpha e^{i(1-k)\theta} + (1 - \beta) \sum_{k=0}^{\infty} g_k^\alpha e^{-i(1-k)\theta} \right\}.$$

For $\theta = \pi$, we have

$$\kappa_C(\pi) = 1 - \frac{1}{2}\mu_\alpha \left\{ (1 + \beta) \sum_{k=0}^{\infty} a_k^\alpha + (1 - \beta) \sum_{k=0}^{\infty} a_k^\alpha \right\}.$$

Hence, from Theorem 4.10 we know that $|\kappa_C(\pi)| \leq 1$ implies $\mu_\alpha \leq 2^{1-\alpha}$. $\qquad\square$

The well-known similar results for the advection diffusion equation are the following.

Theorem 4.12 *Let $\alpha = 2$.*

(i) The upwind scheme is von Neumann stable if, and only if, $v + 2\mu_\alpha \leq 1$.
(ii) The Lax-Wendroff scheme is von Neumann stable if, and only if, $v^2 + 2\mu_\alpha \leq 1$.
(iii) The central scheme is von Neumann stable if, and only if, $v^2 \leq 2\mu_\alpha \leq 1$.

Proof The proof of (i), (ii) and (iii) can be found in various works such as [102, 127, 157] respectively. $\qquad\square$

4.3   Second Order Finite Difference Methods

In Sect. 4.2, we have seen that the accuracy of the numerical methods is affected by the first order approximation of the fractional derivative. Therefore, if we can improve the accuracy of the approximation of the fractional derivative, we will be able to obtain a more accurate numerical method. The numerical methods introduced in this section will be based on a second order approximation of the Riemann-Liouville fractional derivative derived from the integral formula.

4.3.1   Linear Spline Fractional Derivative Approximation

In order to be able to obtain a second order numerical method in space, we need to approximate the fractional derivative by a second order approximation. Although, at the moment there are different approaches in literature to obtain a second order approximation, we present only one of those approaches. This approach has been developed in [130, 140] and it has been based in some of the ideas presented in previous works such as [42, 83, 86].

We start to discuss the approximation of the fractional derivative and, as previously, for clarity purposes, we assume we have a function of one variable $u(x)$. The approximation for the case when we have a function $u(x, t)$, considered in the next sections, is then straightforward.

Let us consider the left Riemann-Liouville fractional derivative, for $\alpha \in (1, 2)$,

$$D_+^\alpha u(x) = \frac{1}{\Gamma(2 - \alpha)} \frac{d^2}{dx^2} \int_{-\infty}^{x} u(\xi)(x - \xi)^{1-\alpha} d\xi. \tag{4.39}$$

We define the mesh points $x_j = j\Delta x$, $j \in \mathbb{Z}$, where Δx denotes the uniform space step. Let us denote

$$I_+^{2-\alpha} u(x) = \frac{1}{\Gamma(2 - \alpha)} \int_{-\infty}^{x} u(\xi)(x - \xi)^{1-\alpha} d\xi. \tag{4.40}$$

First, we explain how to derive the approximation. Later on, we discuss under which conditions this approach is valid.

We start to do the following approximation, at x_j,

$$\frac{d^2}{dx^2} I_+^{2-\alpha} u(x_j) \simeq \frac{1}{(\Delta x)^2} \left[I_+^{2-\alpha} u(x_{j-1}) - 2I_+^{2-\alpha} u(x_j) + I_+^{2-\alpha} u(x_{j+1}) \right].$$

For each x_j we need to calculate $I_+^{2-\alpha} u(x_j)$.

The information from the kernel function is a precious information and therefore this function should never be simplified or approximated. The best thing to do is to approximate the function itself in a way that those approximations together with the discontinuous kernel become an integrable function.

We compute these integrals by approximating $u(\xi)$ by a linear spline $s_j(\xi)$, whose nodes and knots are chosen at $x_k, k = \ldots, j-1, j$. Hence, an approximation to $I_+^{2-\alpha} u(x_j)$ becomes $\tilde{I}_+^{2-\alpha} u(x_j)$ defined by

$$\tilde{I}_+^{2-\alpha} u(x_j) = \frac{1}{\Gamma(2-\alpha)} \int_{-\infty}^{x_j} s_j(\xi)(x_j - \xi)^{1-\alpha} d\xi. \tag{4.41}$$

The spline $s_j(\xi)$ interpolates the points $\{(x_k) : k \le j\}$ and is of the form

$$s_j(\xi) = \sum_{k=-\infty}^{j} u(x_k) s_{j,k}(\xi), \tag{4.42}$$

with $s_{j,k}(\xi)$ defined in each interval $[x_{k-1}, x_{k+1}]$, for $k \le j - 1$, as the hat function and for $k = j$ is only defined in $[x_{j-1}, x_j]$. From (4.41) and (4.42),

$$\tilde{I}_+^{2-\alpha} u(x_j) = \frac{1}{\Gamma(2-\alpha)} \sum_{k=-\infty}^{j} u(x_k) \int_{x_{k-1}}^{x_{k+1}} s_{j,k}(\xi)(x_j - \xi)^{1-\alpha} d\xi$$

$$= \frac{1}{\Gamma(2-\alpha)} \sum_{k=-\infty}^{j} u(x_k) \frac{(\Delta x)^{2-\alpha}}{(2-\alpha)(3-\alpha)} a_{j,k}^{\alpha}, \tag{4.43}$$

where the $a_{j,k}^{\alpha}$ are such that, for $k = j$, we have $a_{j,j}^{\alpha} = 1$ and for $k \le j - 1$,

$$a_{j,k}^{\alpha} = (j - k + 1)^{3-\alpha} - 2(j - k)^{3-\alpha} + (j - k - 1)^{3-\alpha}. \tag{4.44}$$

Therefore,

$$\tilde{I}_+^{2-\alpha} u(x_j) = \frac{(\Delta x)^{2-\alpha}}{\Gamma(4-\alpha)} \sum_{k=-\infty}^{j} u(x_k) a_{j,k}^{\alpha}. \tag{4.45}$$

Hence, an approximation for

$$\frac{d^2}{dx^2} I_+^{2-\alpha} u(x_j)$$

is given by

$$\frac{d^2}{dx^2} I_+^{2-\alpha} u(x_j) \approx \frac{1}{(\Delta x)^2} \left[\tilde{I}_+^{2-\alpha} u(x_{j-1}) - 2\tilde{I}_+^{2-\alpha} u(x_j) + \tilde{I}_+^{2-\alpha} u(x_{j+1}) \right] \quad (4.46)$$

$$= \frac{(\Delta x)^{-\alpha}}{\Gamma(4-\alpha)} \left[\sum_{k=-\infty}^{j-1} u(x_k) a_{j-1,k}^{\alpha} - 2 \sum_{k=-\infty}^{j} u(x_k) a_{j,k}^{\alpha} + \sum_{k=-\infty}^{j+1} u(x_k) a_{j+1,k}^{\alpha} \right].$$

Let us define the following fractional operator

$$\delta_{l,2}^{\alpha} u(x_j) = \frac{1}{\Gamma(4-\alpha)} \left\{ \sum_{k=-\infty}^{j+1} q_{j,k}^{\alpha} u(x_k) \right\}, \quad (4.47)$$

where

$$q_{j,k}^{\alpha} = a_{j-1,k}^{\alpha} - 2a_{j,k}^{\alpha} + a_{j+1,k}^{\alpha}, \ k \leq j-1,$$
$$q_{j,j}^{\alpha} = -2a_{j,j}^{\alpha} + a_{j+1,j}^{\alpha},$$
$$q_{j,j+1}^{\alpha} = a_{j+1,j+1}^{\alpha}. \quad (4.48)$$

Therefore, an approximation of (4.39) can be given by $\dfrac{\delta_{l,2}^{\alpha} u(x_j)}{(\Delta x)^{\alpha}}$. In the notation of the operator $\delta_{l,2}^{\alpha} u$, the l stands for the left Riemann-Liouville fractional derivative and the 2 for the order of accuracy of the approximation that will be discussed bellow.

We can rewrite the fractional operator (4.47) as

$$\delta_{l,2}^{\alpha} u(x_j) = \frac{1}{\Gamma(4-\alpha)} \sum_{m=-1}^{\infty} q_m^{\alpha} u(x_{j-m}), \quad (4.49)$$

where

$$q_m^{\alpha} = a_{m-1}^{\alpha} - 2a_m^{\alpha} + a_{m+1}^{\alpha}, \quad m \geq 1$$
$$q_0^{\alpha} = -2a_0^{\alpha} + a_1^{\alpha},$$
$$q_{-1}^{\alpha} = a_0^{\alpha}, \quad (4.50)$$

with

$$a_0^{\alpha} = 1, \quad a_m^{\alpha} = (m+1)^{3-\alpha} - 2m^{3-\alpha} + (m-1)^{3-\alpha}, \ m \geq 1.$$

The series (4.49) converges absolutely, for each $\alpha \in (1,2)$, and for every bounded function $u(x)$. Note that for $\alpha = 1$ and $\alpha = 2$ the coefficients (4.48) are such that $q_m^\alpha = 0$, for $m > 1$. For $\alpha = 1$, $q_1^\alpha = -1$, $q_0^\alpha = 0$, $q_{-1}^\alpha = 1$ and for $\alpha = 2$, $q_1^\alpha = 1$, $q_0^\alpha = -2$, $q_{-1}^\alpha = 1$.

We now prove this approximation is second order accurate. In the next lemma and theorem, for the sake of clarity, we denote the partial derivative of u in x of order r by $u^{(r)}$. We start with a useful lemma that can be found in [140].

Lemma 4.1 *Let $I_k = [x_{k-1}, x_k]$ and $u \in C^{(4)}(I_k)$. For $\xi \in I_k$ and*

$$s_k(\xi) = \frac{x_k - \xi}{\Delta x} u(x_{k-1}) + \frac{\xi - x_{k-1}}{\Delta x} u(x_k), \tag{4.51}$$

we have

$$u(\xi) - s_k(\xi) = -\sum_{r=2}^{3} \frac{1}{r!} u^{(r)}(\xi) l_{k,r}(\xi) - \frac{1}{4!} u^{(4)}(\eta_k) l_{k,4}(\xi), \quad \eta_k \in I_k,$$

where $|l_{k,r}(\xi)| \leq C_r (\Delta x)^r$, $r = 2, 3, 4$, and C_r are positive constants that do not depend on k and Δx.

Proof For $\xi \in [x_{k-1}, x_k]$,

$$u(\xi) - s_k(\xi) = u(\xi) - \frac{x_k - \xi}{\Delta x} u(x_{k-1}) - \frac{\xi - x_{k-1}}{\Delta x} u(x_k).$$

Using Taylor expansions, we obtain

$$u(\xi) - s_k(\xi) = -\sum_{r=2}^{3} \frac{1}{r!} u^{(r)}(\xi) l_{k,r}(\xi) - \frac{1}{4!} u^{(4)}(\eta_k) l_{k,r}(\xi),$$

where $l_{k,r}(\xi)$ are functions which depend on Δx and x_k, given by

$$l_{k,r}(\xi) = \frac{x_k - \xi}{\Delta x} (x_k - \xi - \Delta x)^r - \frac{\xi - x_k + \Delta x}{\Delta x} (x_k - \xi)^r$$

$$= 2(\Delta x)^{-1}(x_k - \xi)^{r+1} - (x_k - \xi)^r$$

$$+ \sum_{p=0}^{r-1} \binom{r}{p} (x_k - \xi)^{p+1}(-1)^{r-p}(\Delta x)^{r-p-1}. \tag{4.52}$$

It is easy to conclude that $|l_{k,r}(\xi)| \leq C(\Delta x)^r$, for $\xi \in [x_{k-1}, x_k]$. $\square$

We proceed to present the result that shows this approximation is second order accurate for functions defined in the real line. We consider the set of functions that are sufficiently smooth, that is, those that are differentiable up to a certain order and vanish sufficiently rapidly at infinity together with their derivatives. For detailed discussions on other spaces where we can consider the existence of the fractional derivatives, when defined in the real line, see [119, pages 109,139, 148].

In this context the assumption of functions that vanish sufficiently rapidly at infinity together with their derivatives usually means that for each $k = 0, 1, 2, \ldots$ there exists m such that $\sup(1 + |x|)^m |u^{(k)}(x)| < \infty$.

For instance the fractional derivative $D_+^\alpha u$ of order $\alpha \in (0, 1)$ exists for u continuously differentiable and with its derivative $u'(x)$ vanishing at infinity as $|x|^{\alpha-1-\epsilon}$, $\epsilon > 0$ (see Exercise 4.7). Also for the Lévy alpha-stable distributions, introduced briefly in the second chapter, which are fundamental solutions of the fractional diffusion equation, the asymptotic decay of the tail is $\mathcal{O}(|x|^{-(\alpha+1)})$, when $x \to \pm\infty$.

Theorem 4.13 (Order of Accuracy of the Approximation for the Fractional Derivative) *Let $u \in C^{(4)}(\mathbb{R})$, $u^{(4)} \in L_\infty(\mathbb{R})$ and u vanishes sufficiently rapidly at infinity together with their derivatives. We have that*

$$D_+^\alpha u(x_j) - \frac{\delta_l^\alpha u}{(\Delta x)^\alpha}(x_j) = \epsilon_T(x_j),$$

where $|\epsilon_T(x_j)| \leq C(\Delta x)^2$, and C is a constant independent of Δx.

Proof We start to quantify the error related to the approximation of the second order derivative located outside the fractional integral. For $j = 1, \ldots, N - 1$, there exists an $\eta_j \in (x_{j-1}, x_{j+1})$ such that

$$\frac{d^2}{dx^2} I_+^{2-\alpha} u(x_j) = \frac{I_+^{2-\alpha} u(x_{j+1}) - 2 I_+^{2-\alpha} u(x_j) + I_+^{2-\alpha} u(x_{j-1})}{(\Delta x)^2} + \epsilon_1(\eta_j),$$

where

$$\epsilon_1(\eta_j) = -\frac{(\Delta x)^2}{12} \frac{d^4}{dx^4} I_+^{2-\alpha} u(\eta_j) = -\frac{(\Delta x)^2}{12} \frac{d^2}{dx^2} D_+^\alpha u(\eta_j).$$

Considering the composition property between integer and fractional derivatives, see first chapter or [75, p. 89], it follows that

$$|\epsilon_1(\eta_j)| = \frac{(\Delta x)^2}{12} \left| D_+^{\alpha+2} u(\eta_j) \right|,$$

where the fractional derivative of order $\alpha + 2$ is defined according to the definitions presented in the first chapter. The fractional derivative of order $\alpha + 2$ exists for $u \in C^4(\mathbb{R})$ and for functions that vanish sufficiently rapidly at infinity together with their derivatives (see Exercises 4.7 and 4.8). Therefore $\epsilon_1(x_j) = \mathcal{O}((\Delta x)^2)$.

Let us define the error $E_S(x_j)$, such that,

$$I_+^{2-\alpha}u(x_{j-1}) - 2I_+^{2-\alpha}u(x_j) + I_+^{2-\alpha}u(x_{j+1})$$
$$= \tilde{I}_+^{2-\alpha}u(x_{j-1}) - 2\tilde{I}_+^{2-\alpha}u(x_j) + \tilde{I}_+^{2-\alpha}u(x_{j+1}) + E_S(x_j).$$

We have

$$D_+^\alpha u(x_j) = \frac{1}{(\Delta x)^2}\left[\tilde{I}_+^{2-\alpha}u(x_{j-1}) - 2\tilde{I}_+^{2-\alpha}u(x_j) + \tilde{I}_+^{2-\alpha}u(x_{j+1})\right]$$
$$+ \frac{1}{(\Delta x)^2}E_S(x_j) + \epsilon_1(\eta_j).$$

Hence,

$$D_+^\alpha u(x_j) = \frac{\delta_l^\alpha u}{(\Delta x)^\alpha}(x_j) + \epsilon_1(\eta_j) + \frac{1}{(\Delta x)^2}E_S(x_j).$$

We are now going to compute the error $E_S(x_j)$. We have

$$E_S(x_j) = \frac{1}{\Gamma(2-\alpha)}\sum_{k=-\infty}^{j-1}\int_{x_{k-1}}^{x_k}(u(\xi) - s_{j-1,k}(\xi))(x_{j-1} - \xi)^{1-\alpha}d\xi$$

$$-\frac{2}{\Gamma(2-\alpha)}\sum_{k=-\infty}^{j}\int_{x_{k-1}}^{x_k}(u(\xi) - s_{j,k}(\xi))(x_j - \xi)^{1-\alpha}d\xi$$

$$+\frac{1}{\Gamma(2-\alpha)}\sum_{k=-\infty}^{j+1}\int_{x_{k-1}}^{x_k}(u(\xi) - s_{j+1,k}(\xi))(x_{j+1} - \xi)^{1-\alpha}d\xi.$$

Taking in consideration the previous lemma, let us denote

$$E_S(x_j) = -\frac{1}{\Gamma(2-\alpha)}\sum_{r=2}^{4}\frac{1}{r!}E_r(x_j), \tag{4.53}$$

where $E_r(x_j)$ are defined as follows. For $r = 2$ and $r = 3$,

$$E_r(x_j) = \sum_{k=-\infty}^{j-1}\int_{x_{k-1}}^{x_k}l_{k,r}(\xi)u^{(r)}(\xi)(x_{j-1} - \xi)^{1-\alpha}d\xi$$

$$-2\sum_{k=-\infty}^{j}\int_{x_{k-1}}^{x_k}l_{k,r}(\xi)u^{(r)}(\xi)(x_j - \xi)^{1-\alpha}d\xi$$

$$+\sum_{k=-\infty}^{j+1}\int_{x_{k-1}}^{x_k}l_{k,r}(\xi)u^{(r)}(\xi)(x_{j+1} - \xi)^{1-\alpha}d\xi \tag{4.54}$$

and for $r = 4$

$$E_r(x_j) = \sum_{k=-\infty}^{j-1} u^{(4)}(\eta_k) \int_{x_{k-1}}^{x_k} l_{k,r}(\xi)(x_{j-1} - \xi)^{1-\alpha} d\xi$$

$$-2 \sum_{k=-\infty}^{j} u^{(4)}(\eta_k) \int_{x_{k-1}}^{x_k} l_{k,r}(\xi)(x_j - \xi)^{1-\alpha} d\xi$$

$$+ \sum_{k=-\infty}^{j+1} u^{(4)}(\eta_k) \int_{x_{k-1}}^{x_k} l_{k,r}(\xi)(x_{j+1} - \xi)^{1-\alpha} d\xi. \qquad (4.55)$$

For $r = 2, 3$, by a change of variables, we obtain

$$E_r(x_j) = \sum_{k=-\infty}^{j} \int_{x_{k-1}}^{x_k} l_{k,r}(\xi) \left[u^{(r)}(\xi + \Delta x) - 2u^{(r)}(\xi) \right.$$

$$\left. + u^{(r)}(\xi - \Delta x) \right] (x_j - \xi)^{1-\alpha} d\xi.$$

By hypothesis u vanishes sufficiently rapidly at infinity together with their derivatives. Therefore, there exists a $x_{-N} = -N\Delta x$ such that

$$\int_{-\infty}^{x_{-N}} u^{(4)}(\xi)(x_j - \xi)^{1-\alpha} d\xi < \mathcal{O}((\Delta x)^4).$$

Hence,

$$E_r(x_j) = \sum_{k=-N+1}^{j} \int_{x_{k-1}}^{x_k} l_{k,r}(\xi) \left[u^{(r)}(\xi + \Delta x) \right.$$

$$\left. - 2u^{(r)}(\xi) + u^{(r)}(\xi - \Delta x) \right] (x_j - \xi)^{1-\alpha} d\xi$$

$$+ \mathcal{O}((\Delta x)^4).$$

For $r = 2$, there exists $\xi_k \in [x_{k-1}, x_k]$ such that

$$E_2(x_j) = \frac{(\Delta x)^2}{2} \sum_{k=-N+1}^{j} u^{(4)}(\xi_k) c_{j,k,2} + \mathcal{O}((\Delta x)^4),$$

where

$$c_{j,k,2} = \int_{x_{k-1}}^{x_k} l_{k,r}(\xi)(x_j - \xi)^{1-\alpha} d\xi.$$

Since, by Lemma 4.1,

$$|c_{j,k,2}| \leq (\Delta x)^2 \int_{x_{k-1}}^{x_k} (x_j - \xi)^{1-\alpha} d\xi$$

and

$$\int_{x_a}^{x_j} (x_j - \xi)^{1-\alpha} d\xi = \frac{1}{2-\alpha}(x_j - x_a)^{2-\alpha},$$

we have

$$|E_2(x_j)| \leq \frac{(\Delta x)^4}{2(2-\alpha)}||u^{(4)}||_\infty (x_j - x_a)^{2-\alpha} + \mathcal{O}((\Delta x)^4). \qquad (4.56)$$

For $r = 3$, there exists $\xi_{k_1}, \xi_{k_2} \in [x_{k-1}, x_k]$, such that

$$E_3(x_j) = \sum_{k=-N+1}^{j} \Delta x (u^{(4)}(\xi_{k_1}) - u^{(4)}(\xi_{k_2}))c_{j,k,3} + \mathcal{O}((\Delta x)^4)$$

and

$$|c_{j,k,3}| \leq (\Delta x)^3 \int_{x_{k-1}}^{x_k} (x_j - \xi)^{1-\alpha} d\xi.$$

We have

$$|E_3(x_j)| \leq \frac{2(\Delta x)^4}{(2-\alpha)}||u^{(4)}||_\infty (x_j - x_a)^{2-\alpha} + \mathcal{O}((\Delta x)^4). \qquad (4.57)$$

Finally for $r = 4$, we bound each integral of (4.55) separately. For the first integral we have

$$\sum_{k=-N+1}^{j-1} u^{(4)}(\eta_k) \int_{x_{k-1}}^{x_k} l_{k,4}(\xi)(x_{j-1} - \xi)^{1-\alpha} d\xi$$

$$\leq (\Delta x)^4 ||u^{(4)}||_\infty \sum_{k=N_a+1}^{j-1} \int_{x_{k-1}}^{x_k} (x_{j-1} - \xi)^{1-\alpha} d\xi$$

$$= \frac{(\Delta x)^4}{2-\alpha}||u^{(4)}||_\infty (x_{j-1} - x_a)^{2-\alpha}.$$

Therefore, since $(a+b)^p \leq |a|^p + |b|^p$, for $0 < p \leq 1$, we have

$$\sum_{k=-N+1}^{j-1} u^{(4)}(\eta_k) \int_{x_{k-1}}^{x_k} l_{k,4}(\xi)(x_{j-1} - \xi)^{1-\alpha} d\xi$$

$$\leq \frac{(\Delta x)^4}{2-\alpha} ||u^{(4)}||_\infty ((x_j - x_a)^{2-\alpha} + (\Delta x)^{2-\alpha}).$$

Similarly, for the second integral we have

$$\sum_{k=-N+1}^{j} u^{(4)}(\eta_k) \int_{x_{k-1}}^{x_k} l_{k,4}(\xi)(x_j - \xi)^{1-\alpha} d\xi \leq \frac{(\Delta x)^4}{2-\alpha} ||u^{(4)}||_\infty (x_j - x_a)^{2-\alpha}$$

and for the third integral

$$\sum_{k=-N+1}^{j+1} u^{(4)}(\eta_k) \int_{x_{k-1}}^{x_k} l_{k,4}(\xi)(x_{j+1} - \xi)^{1-\alpha} d\xi$$

$$\leq \frac{(\Delta x)^4}{2-\alpha} ||u^{(4)}||_\infty ((x_j - x_a)^{2-\alpha} + (\Delta x)^{2-\alpha}).$$

Finally, we have

$$|E_4(x_j)| \leq \frac{3(\Delta x)^4}{2-\alpha} ||u^{(4)}||_\infty (x_j - x_a)^{2-\alpha} + \frac{2(\Delta x)^{6-\alpha}}{2-\alpha} ||u^{(4)}||_\infty + \mathscr{O}((\Delta x)^4). \tag{4.58}$$

From (4.56), (4.57) and (4.58) we can conclude that the error $E_S(x_j)$ defined by (4.53) is of order $\mathscr{O}((\Delta x)^4)$ and therefore the $\epsilon_2(x_j)$ is of order $\mathscr{O}((\Delta x)^2)$. $\qquad\square$

For the right Riemann-Liouville fractional derivative

$$D_-^\alpha u(x) = \frac{1}{\Gamma(2-\alpha)} \frac{d^2}{dx^2} \int_x^\infty u(\xi)(\xi - x)^{1-\alpha} d\xi, \quad \alpha \in (1,2), \tag{4.59}$$

an approximation can be given by

$$\frac{\delta_{r,2}^\alpha u(x_j)}{(\Delta x)^\alpha},$$

where

$$\delta_{r,2}^\alpha u(x_j) = \frac{1}{\Gamma(4-\alpha)} \sum_{m=-1}^\infty q_m^\alpha u(x_{j+m}). \tag{4.60}$$

In the notation $\delta^\alpha_{r,2} u$ of the operator, the r stands for the right Riemann-Liouville fractional derivative and 2 stands for the order of accuracy of the approximation. A similar approach, to the one presented for the left Riemann-Liouville fractional derivative, can be done to prove that this approximation is second order accurate.

4.3.2   Implicit Numerical Methods

We have presented a discretization of the fractional derivative that is second order accurate. Therefore, we are now able to derive a second order accurate numerical method, if we consider a second order discretization in time.

We start to discuss a numerical method for Eq. (4.5), when $V = 0$, $\beta = 1$ and with a source term $f(x, t)$, that is,

$$\frac{\partial u}{\partial t} = D \frac{\partial^\alpha u}{\partial x^\alpha} + f(x, t). \tag{4.61}$$

Let us consider the time weighted average discretization. We consider $t \in [0, T]$ and the time discretization $t_n = n \Delta t$. For the uniform space step Δx and time step Δt, let

$$\mu_\alpha = \frac{D \Delta t}{(\Delta x)^\alpha}.$$

From Eq. (4.61) we can obtain the explicit Euler and implicit Euler numerical methods, respectively

$$\frac{U_j^{n+1} - U_j^n}{\Delta t} = \frac{D}{(\Delta x)^\alpha} \delta^\alpha_{l,2} U_j^n + f_j^n, \tag{4.62}$$

$$\frac{U_j^{n+1} - U_j^n}{\Delta t} = \frac{D}{(\Delta x)^\alpha} \delta^\alpha_{l,2} U_j^{n+1} + f_j^{n+1}, \tag{4.63}$$

where the operator $\delta^\alpha_{l,2}$ is defined by (4.49). We can also arrive at the weighted average θ-scheme,

$$U_j^{n+1} - U_j^n = \mu_j^\alpha \left\{ (1 - \theta) \delta^\alpha_{l,2} U_j^n + \theta \delta^\alpha_{l,2} U_j^{n+1} \right\} + \theta \Delta t f_j^{n+1} + (1 - \theta) \Delta t f_j^n, \tag{4.64}$$

where $\theta \in [1/2, 1]$.

Note that for $\alpha = 2$, the operator (4.47) is the central second order operator $\delta^2 U_j^n$

$$\delta^2 U_j^n = U_{j+1}^n - 2U_j^n + U_{j-1}^n.$$

We have the following numerical method

$$\left(1 - \theta\mu\delta_{l,2}^\alpha\right) U_j^{n+1} = \left(1 + (1-\theta)\mu\delta_{l,2}^\alpha\right) U_j^n + \Delta t f_j^{n+\theta}, \tag{4.65}$$

where

$$f_j^{n+\theta} = \theta f_j^n + (1-\theta) f_j^{n+1}.$$

Theorem 4.14 *Let u be a solution of the fractional partial differential equation and satisfying the conditions of the previous theorem, that is, $u(\cdot, t) \in C^{(4)}(\mathbb{R})$, $u^{(4)}(\cdot, t) \in L_\infty(\mathbb{R})$ and u vanishes sufficiently rapidly at infinity together with their derivatives in the x direction. Additionally $u(x, \cdot) \in C^{(3)}(0, T)$.*

The truncation error of the weighted numerical method (4.65) is of order $\mathcal{O}((\Delta x)^2) + \mathcal{O}((\Delta t)^{m_\theta})$, where $m_\theta = 1$, for $\theta \in (1/2, 1]$ and $m_\theta = 2$, for $\theta = 1/2$.

Proof We denote the exact solution at the discrete points (x_j, t_n) as $u_j^n := u(x_j, t_n)$. The truncation error for the numerical method (4.65) is given by

$$T_j^n = \frac{u_j^{n+1} - u_j^n}{\Delta t} - \frac{D}{(\Delta x)^\alpha}\left(\theta\delta_{l,2}^\alpha u_j^{n+1} + (1-\theta)\delta_{l,2}^\alpha u_j^n\right) - f_j^{n+\theta}.$$

We have that

$$\frac{u_j^{n+1} - u_j^n}{\Delta t} = \frac{\partial u(x_j, t_n)}{\partial t} + \frac{\Delta t}{2}\frac{\partial^2 u(x_j, t_n)}{\partial t^2} + \mathcal{O}((\Delta t)^2). \tag{4.66}$$

Using the previous theorem we have

$$\begin{aligned}
T_j^n = &\frac{\partial u(x_j, t_n)}{\partial t} + \frac{\Delta t}{2}\frac{\partial^2 u(x_j, t_n)}{\partial t^2} + \mathcal{O}((\Delta t)^2) \\
&- \theta\left(d_j\frac{\partial^\alpha u(x_j, t_{n+1})}{\partial x^\alpha} + \mathcal{O}((\Delta x)^2)\right) \\
&- (1-\theta)\left(D\frac{\partial^\alpha u(x_j, t_n)}{\partial x^\alpha} + \mathcal{O}((\Delta x)^2)\right) - f_j^{n+\theta}.
\end{aligned}$$

Therefore,

$$\begin{aligned}
T_j^n = &\frac{\partial u(x_j, t_n)}{\partial t} + \frac{\Delta t}{2}\frac{\partial^2 u(x_j, t_n)}{\partial t^2} + -(1-\theta)\frac{\partial u(x_j, t_n)}{\partial t} - \theta\frac{\partial u(x_j, t_{n+1})}{\partial t} \\
&+ \mathcal{O}((\Delta t)^2) + \mathcal{O}((\Delta x)^2).
\end{aligned}$$

Finally,

$$T_j^n = (\frac{1}{2} - \theta)\Delta t \frac{\partial^2 u(x_j, t_n)}{\partial t^2} + \mathcal{O}((\Delta t)^2) + \mathcal{O}((\Delta x)^2).$$

$\square$

In order to derive stability conditions for the finite difference schemes, we apply the von Neumann analysis as done previously for the first order numerical methods. Considering a single mode $\kappa^n e^{ij\phi}$, with $\phi = \xi \Delta x$, its time evolution is determined by the numerical scheme. The stability conditions will be satisfied if the amplitude factor κ does not grow in time, that is, if we have $|\kappa(\phi)| \le 1$, for all $\phi \in [-\pi, \pi]$. We call it now ϕ instead of θ as in the previous chapters, since we use the θ for the weight of the method.

Before presenting the stability analysis, we present the properties of the coefficients q_m^α.

Lemma 4.2 *Consider the coefficients q_m^α defined by (4.50), for $\alpha \in (1, 2]$. Then*

(a) $q_{-1}^\alpha = 1$, $q_0^\alpha \le 0$, $q_m^\alpha \ge 0$, $m \ge 2$, $\lim\limits_{m \to \infty} q_m^\alpha = 0$ *and* $q_{m+1}^\alpha \le q_m^\alpha \le q_2^\alpha$.

(b) $\displaystyle\sum_{m=2}^{\infty} q_m^\alpha = -3 + 3 \times 2^{3-\alpha} - 3^{3-\alpha}.$

(c) $\displaystyle\sum_{m=-1}^{\infty} q_m^\alpha = 0.$

Proof

(a) By computing the first three coefficients we arrive at the equalities

$$q_{-1}^\alpha = 1, \quad q_0^\alpha = 2^{3-\alpha} - 4, \quad q_1^\alpha = 3^{3-\alpha} - 4 \times 2^{3-\alpha} + 6.$$

The coefficient q_0^α is negative for all $\alpha \in (1, 2]$ and the coefficient q_1^α can be positive or negative depending on the value of α.

The q_m^α, $m \ge 2$, are of the form

$$q_m^\alpha = (m + 2)^{3-\alpha} - 4(m + 1)^{3-\alpha} + 6m^{3-\alpha} - 4(m - 1)^{3-\alpha} + (m - 2)^{3-\alpha}.$$

Hence,

$$q_m^\alpha = m^{3-\alpha}\left[\left(1 + \frac{2}{m}\right)^{3-\alpha} - 4\left(1 + \frac{1}{m}\right)^{3-\alpha}\right.$$
$$\left. + 6 - 4\left(1 - \frac{1}{m}\right)^{3-\alpha} + \left(1 - \frac{2}{m}\right)^{3-\alpha}\right]$$
$$= m^{3-\alpha}\left[\sum_{k=0}^{\infty}\binom{3-\alpha}{k}\left(\frac{2}{m}\right)^k\right.$$

$$-4\sum_{k=0}^{\infty}\binom{3-\alpha}{k}\left(\frac{1}{m}\right)^{k}+6-4\sum_{k=0}^{\infty}\binom{3-\alpha}{k}\left(\frac{-1}{m}\right)^{k}$$
$$+\sum_{k=0}^{\infty}\binom{3-\alpha}{k}\left(\frac{-2}{m}\right)^{k}\Bigg].$$

Therefore,

$$q_m^{\alpha}=m^{3-\alpha}\left[\sum_{k=4}^{\infty}\binom{3-\alpha}{k}\left(\frac{2}{m}\right)^{k}-4\sum_{k=4}^{\infty}\binom{3-\alpha}{k}\left(\frac{1}{m}\right)^{k}\right.$$
$$\left.-4\sum_{k=4}^{\infty}\binom{3-\alpha}{k}\left(\frac{-1}{m}\right)^{k}+\sum_{k=4}^{\infty}\binom{3-\alpha}{k}\left(\frac{-2}{m}\right)^{k}\right]$$
$$=m^{3-\alpha}\left[\frac{(3-\alpha)(3-\alpha-1)(3-\alpha-2)(3-\alpha-3)}{4!}\frac{24}{m^{4}}+\cdots\right]$$
$$=\frac{1}{m^{\alpha-1}}\left[\frac{(3-\alpha)(2-\alpha)(1-\alpha)(-\alpha)}{4!}\frac{24}{m^{2}}+\cdots\right].\tag{4.67}$$

Considering (4.67) and noting that the k odd terms of the series cancel, the properties
(a) can be easily obtained.

(b) In order to compute the series, let us first compute the sum of the first $M-1$
terms. We have

$$\sum_{m=2}^{M}q_m^{\alpha}=-3+3\times 2^{3-\alpha}-3^{3-\alpha}+s_M,$$

where

$$s_M=-(M-1)^{3-\alpha}+3M^{3-\alpha}-3(M+1)^{3-\alpha}+(M+2)^{3-\alpha}.$$

Similar to what has been done in (a) we can write

$$s_M=M^{3-\alpha}\left[\left(1+\frac{2}{M}\right)^{3-\alpha}-3\left(1+\frac{1}{M}\right)^{3-\alpha}+3-\left(1-\frac{1}{M}\right)^{3-\alpha}\right]$$
$$=M^{3-\alpha}\left[\sum_{k=0}^{\infty}\binom{3-\alpha}{k}\left(\frac{2}{M}\right)^{k}-3\sum_{k=0}^{\infty}\binom{3-\alpha}{k}\left(\frac{1}{M}\right)^{k}+3\right.$$
$$\left.-\sum_{k=0}^{\infty}\binom{3-\alpha}{k}\left(\frac{-1}{M}\right)^{k}\right].$$

Therefore,

$$s_M = M^{3-\alpha} \left[\sum_{k=3}^{\infty} \binom{3-\alpha}{k} \left(\frac{2}{M}\right)^k - 3 \sum_{k=3}^{\infty} \binom{3-\alpha}{k} \left(\frac{1}{M}\right)^k \right.$$

$$\left. - \sum_{k=3}^{\infty} \binom{3-\alpha}{k} \left(\frac{-1}{M}\right)^k \right]$$

$$= M^{3-\alpha} \left[\frac{(3-\alpha)(2-\alpha)(1-\alpha)}{3!} \frac{6}{M^3} + \cdots \right]$$

$$= \frac{1}{M^{\alpha-1}} \left[\frac{(3-\alpha)(2-\alpha)(1-\alpha)}{3!} \frac{6}{M} + \cdots \right]. \tag{4.68}$$

Clearly, we can conclude that

$$\lim_{M \to \infty} s_M = 0.$$

Hence,

$$\sum_{m=2}^{\infty} q_m^{\alpha} = \lim_{M \to \infty} \sum_{m=2}^{M} q_m = -3 + 3 \times 2^{3-\alpha} - 3^{3-\alpha}.$$

(c) This result comes immediately from (b) and from the fact that

$$q_{-1}^{\alpha} + q_0^{\alpha} + q_1^{\alpha} = 3 - 3 \times 2^{3-\alpha} + 3^{3-\alpha}.$$

$$\square$$

Remark 4.2 Note that the previous result on the convergence of the series with the general term q_m^{α} allow us to conclude that the series, defining the operator (4.49), converges absolutely when we have a bounded function u.

The next theorem states the method is unconditionally stable, for $\theta \in [1/2, 1]$.

Theorem 4.15 *The weighted numerical method (4.65) is unconditionally von Neumann stable, for $\theta \in [1/2, 1]$.*

Proof Let us insert the mode $\kappa^n e^{ij\phi}$ into (4.65). We obtain the following

$$\kappa^{n+1}(\phi) \left[e^{ij\phi} - \theta \frac{\mu^{\alpha}}{\Gamma(4-\alpha)} \sum_{m=-1}^{\infty} q_m^{\alpha} e^{i(j-m)\phi} \right]$$

$$= \kappa^n(\phi) \left[e^{ij\phi} + (1-\theta) \frac{\mu^{\alpha}}{\Gamma(4-\alpha)} \sum_{m=-1}^{\infty} q_m^{\alpha} e^{i(j-m)\phi} \right].$$

The amplification factor is given by

$$\kappa(\phi)\left[1 - \theta\frac{\mu^\alpha}{\Gamma(4-\alpha)}\sum_{m=-1}^{\infty}q_m^\alpha e^{-im\phi}\right] = \left[1 + (1-\theta)\frac{\mu^\alpha}{\Gamma(4-\alpha)}\sum_{m=-1}^{\infty}q_m^\alpha e^{-im\phi}\right].$$

Therefore $|\kappa(\phi)| \leq 1$ if and only if the real part of the series is negative, that is,

$$\sum_{m=-1}^{\infty}q_m^\alpha\cos(m\phi) \leq 0,$$

since the imaginary part of the right hand side is smaller, for $\theta \in [1/2, 1]$, because $\theta \geq 1 - \theta$. We can write

$$\sum_{m=-1}^{\infty}q_m^\alpha\cos(m\phi) = (q_{-1}^\alpha + q_1^\alpha)\cos(\phi) + q_0^\alpha + \sum_{m=2}^{\infty}q_m^\alpha\cos(m\phi). \quad (4.69)$$

Since $q_{-1}^\alpha + q_1^\alpha \geq 0$, and $q_m^\alpha \geq 0$, for $m \geq 2$,

$$\sum_{m=-1}^{\infty}q_m\cos(m\phi) \leq (q_{-1}^\alpha + q_1^\alpha) + q_0^\alpha + \sum_{m=2}^{\infty}q_m^\alpha. \quad (4.70)$$

Using Lemma 2. (c), we obtain

$$\sum_{m=-1}^{\infty}q_m^\alpha\cos(m\phi) \leq 0. \quad (4.71)$$

$$\square$$

In what follows, we describe the matricial form of the numerical method, taking in consideration that to implement the numerical method we need to have a computational bounded domain. Therefore, since the function is zero when x goes to infinity we can assume the function is zero outside a certain domain.

The numerical method can be written in the matricial form

$$\left(I - \theta\frac{\mu_\alpha}{\Gamma(4-\alpha)}Q\right)\mathbf{U}^{n+1} = \left(I + (1-\theta)\frac{\mu_\alpha}{\Gamma(4-\alpha)}Q\right)\mathbf{U}^n$$

$$+ \frac{\mu_\alpha}{\Gamma(4-\alpha)}\left(\theta\mathbf{b}^{n+1} + (1-\theta)\mathbf{b}^n\right) + \mathbf{f}^{n+\theta},$$

$$(4.72)$$

where

$$\mathbf{f}^{n+\theta} = \left[\Delta t \theta f_1^{n+1} + (1-\theta) p_1^n \ \ldots \ \Delta t \theta f_{N-1}^{n+1} + (1-\theta) f_{N-1}^n \right]^T,$$

$$\mathbf{U}^n = \left[U_1^n \ \ldots \ U_{N-1}^n \right]^T,$$

$\mathbf{b}^n$ contains the boundary values, μ_α is a diagonal matrix with entries μ_α and Q is related to the fractional operator. The matrix $Q = [Q_{j,k}]$ has the following structure

$$Q_{j,k} = \begin{cases} q_{j-k}^\alpha, & 1 \le k \le j-1 \\ q_0^\alpha, & k = j \\ q_{-1}^\alpha, & k = j+1 \\ 0, & k > j+1. \end{cases}$$

The vector $\mathbf{b}^n$ is given by

$$b_j^n = \begin{cases} 0, & j = 1, \ldots, N-2 \\ q_{-1}^\alpha U_N^n, & j = N-1. \end{cases}$$

assuming that $U_0^n = 0$ and $U_N^n = 0$.

We have discussed the simplest case, when $V = 0$ and $\beta = 1$. However, a similar discussion can be done for the θ-method developed to approximate the more general equation. In order to do that we can consider a similar approximation for the right Riemann-Liouville fractional derivative. The von Neumann stability analysis is similar to what has been done in Sect. 4.2 for the first order approximation. This generalisation is left as an exercise (see Exercise 4.10).

4.3.3 Explicit Numerical Methods

Some of the second order numerical methods relying on implicit discretizations are inadequate for advection dominated flows. In this section, we present a second order numerical method adequate for advection dominated flows, although the derivation of the previous approximation, which main ideas have been developed in [130, 140], allowed the appearance in literature of various second order numerical methods by various authors. The first order numerical method presented in the beginning of this chapter and named of Lax-Wendroff type can now be improved, to be a second order method with the use of this second order approximation.

Consider the more general form of the fractional advection diffusion equation

$$\frac{\partial u}{\partial t}(x,t) + V \frac{\partial u}{\partial x}(x,t) = D \nabla_\beta^\alpha u(x,t), \quad \alpha \in (1,2). \tag{4.73}$$

The approximations for the left and right Riemann-Liouville fractional derivatives are respectively given by

$$\frac{\delta_{l,2}^{\alpha} u(x_j, t)}{(\Delta x)^{\alpha}}, \qquad \frac{\delta_{r,2}^{\alpha} u(x_j, t)}{(\Delta x)^{\alpha}},$$

where the fractional operators are defined, according to Sect. 4.3.1, in the following way,

$$\delta_{l,2}^{\alpha} u(x_j, t) = \frac{1}{\Gamma(4 - \alpha)} \sum_{m=-1}^{\infty} q_m^{\alpha} u(x_{j-m}, t), \qquad (4.74)$$

$$\delta_{r,2}^{\alpha} u(x_j, t) = \frac{1}{\Gamma(4 - \alpha)} \sum_{m=-1}^{\infty} q_m^{\alpha} u(x_{j+m}, t). \qquad (4.75)$$

We consider the operators upwind (4.20) and central (4.21) and the following second order fractional operator

$$\delta_{\beta,2}^{\alpha} u(x_j, t_n) = \frac{1}{2}(1 + \beta)\delta_{l,2}^{\alpha} u(x_j, t_n) + \frac{1}{2}(1 - \beta)\delta_{r,2}^{\alpha} u(x_j, t_n). \qquad (4.76)$$

If we discretize the first derivative with the central difference operator, the second order derivative with second order difference operator and the fractional derivative with the respective fractional difference operator, we have (see the numerical method (4.35) derived previously during this chapter)

$$U_j^{n+1} = U_j^n - \nu \Delta_0 U_j^n + \mu_\alpha \delta_{\beta,2}^{\alpha} U_j^n + \frac{1}{2}\nu^2 \delta^2 U_j^n. \qquad (4.77)$$

Regarding the stability analysis we rely on the Fourier analysis framework as explained previously. The next theorem gives a necessary stability condition, for all $\beta \in [-1, 1]$, that can be found in [131].

Theorem 4.16 *For $\beta \in [-1, 1]$, a necessary von Neumann stability condition for the numerical scheme (4.77) is given by*

$$\nu^2 - \mu_\alpha \frac{1}{2\Gamma(4 - \alpha)} \sum_{m=-1}^{\infty} (-1)^m q_m^{\alpha} \leq 1, \quad \alpha \in (1, 2]. \qquad (4.78)$$

Proof The amplification factor, for the Lax-Wendroff scheme, is given by

$$\kappa(\theta) = 1 - \frac{\nu}{2}(e^{i\theta} - e^{-i\theta}) + \frac{\nu^2}{2}(e^{i\theta} - 2 + e^{-i\theta})$$

$$+ \frac{\mu_\alpha}{2\Gamma(4 - \alpha)} \left((1 + \beta) \sum_{m=-1}^{\infty} q_m^{\alpha} e^{-im\theta} + (1 - \beta) \sum_{m=-1}^{\infty} q_m^{\alpha} e^{im\theta} \right).$$

This can be written as

$$\kappa(\theta) = 1 - iv\sin\theta + v^2(\cos\theta - 1)$$
$$+ \frac{\mu_\alpha}{\Gamma(4-\alpha)}\left(\sum_{m=-1}^{\infty} q_m^\alpha \cos(m\theta) - i\beta \sum_{m=-1}^{\infty} q_m^\alpha \sin(m\theta)\right).$$

For $\theta = \pi$, we have

$$\kappa(\pi) = 1 - 2v^2 + \frac{\mu_\alpha}{2\Gamma(4-\alpha)}\sum_{m=-1}^{\infty} q_m^\alpha \cos(m\theta). \tag{4.79}$$

Therefore,

$$\left| 1 - 2v^2 + \frac{\mu_\alpha}{\Gamma(4-\alpha)}\sum_{m=-1}^{\infty} q_m^\alpha \cos(m\theta) \right| \le 1$$

and this implies the necessary condition (4.78). $\square$

A numerical method of order higher than two suitable for primarily advective flows can be found in [135]. This numerical method is explicit and has the particularity that for $\alpha = 2$ matches the numerical method introduced in [81], and known as QUICKEST. Although it has been introduced a long time ago, the numerical method has been very popular until today. This method, for $\alpha = 2$, had the goal of providing an accurate solution without strong oscillations presented in some higher order methods. It has also been shown to be more efficient than other schemes for highly advective flows, since it makes a reasonable compromise between improved performance and computational cost [128, 158, 161]. The numerical method in [135], that has been adapted for the models described by fractional operators, keeps those properties. One interesting detail that confirms that the method was well designed for these purposes is to observe that the stability regions for values $\alpha < 1.5$ become smaller as discussed in more detail in [135]. Our understanding, concerning this aspect, is that as α goes to one, the fractional operator behaviour is more closer to an advective behaviour than to a diffusive behaviour and therefore there is an interplay between V and D regarding the advective flow. Since the numerical method was designed for advective flows, with $V > 0$, the factor D on the other side of the equation pushes the flow in the opposite direction.

4.4   Exercises and Solutions

Exercises

4.1 Determine for which n and m, the following sentence is true: For $\alpha \in (1, 2)$, if $u \in C^n(\mathbb{R})$ and $u^{(k)} \in L_1(\mathbb{R})$, $k = 0, 1, \ldots, m$, the following integral exists

$$\int_{-\infty}^{\infty} |\xi|^{\alpha+1} |\mathscr{F}(u)(\xi)| d\xi.$$

4.2 Prove that the implicit Euler method, for the diffusion equation (4.7), given by

$$\frac{U_j^{n+1} - U_j^n}{\Delta t} = \frac{D}{(\Delta x)^\alpha} \sum_{k=0}^{\infty} g_k^\alpha U_{j-k}^{n+1}$$

is consistent with order of accuracy of $\mathscr{O}(\Delta x) + \mathscr{O}(\Delta t)$.

4.3 Let $\alpha > 0$ and

$$g_k^\alpha = (-1)^k \binom{\alpha}{k}, \quad k \in \mathbb{N}_0.$$

Prove that:

(a) The coefficients can be obtained by the recurrence formula

$$g_0^\alpha = 1, \qquad g_{k+1}^\alpha = -\frac{\alpha - k}{k + 1} g_k^\alpha, \quad k \geq 1.$$

(b) $\displaystyle\sum_{k=0}^{\infty} (-1)^k g_k^\alpha = 2^\alpha.$

4.4 Prove that the explicit Euler numerical method

$$\frac{U_j^{n+1} - U_j^n}{\Delta t} = \frac{D}{(\Delta x)^\alpha} \sum_{k=0}^{\infty} g_k^\alpha U_{j-k}^n$$

is unstable.

4.5 Prove that the Lax-Wendroff type scheme derived for the fractional advection equation (4.5) with a source term on the right hand side $p(x, t)$, is given by

$$U_j^{n+1} = U_j^n - v\Delta_0 U_j^n + \mu_\alpha \delta_\beta^\alpha U_j^n + \frac{1}{2} v^2 \delta^2 U_j^n + \Delta t\, \tilde{p}(x_j, t_n),$$

where

$$\tilde{p}(x_j, t_n) = p(x_j, t_n) + \frac{1}{2}\Delta t(-Vp_x(x_j, t_n) + p_t(x_j, t_n)).$$

Here, p_t and p_x are the partial derivatives of p. If we do not know the source term and only know a discrete set of values, justify that a good choice for evaluate $\tilde{p}(x, t)$ is

$$\tilde{p}(x_j, t_n) \approx p_j^n + \frac{1}{2}\Delta t\left(-V\frac{p_{j+1}^n - p_{j-1}^n}{2\Delta x} + \frac{p_j^{n+1} - p_j^n}{\Delta t}\right)$$

$$= \frac{1}{2}\left(p_j^{n+1} + p_j^n\right) - \frac{v}{4}\left(p_{j+1}^n - p_{j-1}^n\right),$$

where $p_j^n = p(x_j, t_n)$.

4.6 Derive a second order approximation for the right Riemann-Liouville fractional derivative

$$D_-^\alpha u(x) = \frac{1}{\Gamma(2-\alpha)}\frac{d^2}{dx^2}\int_x^\infty u(\xi)(\xi - x)^{1-\alpha}d\xi.$$

4.7 Let u be a continuously differentiable function and with its derivative u' vanishing at infinity as $|x|^{\alpha-1-\epsilon}$, $\epsilon > 0$. Prove that the fractional derivative $D_+^\alpha u$ exists.

4.8 Let $\alpha \in (0, 1)$. Prove that for a sufficiently good function, then

$$\frac{d^m}{dx^m}\left[D_+^\alpha u(x)\right] = D_+^{\alpha+m} u(x).$$

4.9 Let q_m^α be the coefficients defined by (4.50). Prove that

$$\sum_{m=-1}^\infty q_m^\alpha \cos(m\phi) \leq 0.$$

4.10 (a) Derive the Crank-Nicolson method for the general fractional advection diffusion equation, defined in the real line, and using the second order approximation.

(b) Using the von Neumann analysis, prove that the numerical method is stable.

Solutions

4.1 If

$$|\mathcal{F}(u)(\xi)| \le C|\xi|^{-4}$$

then

$$\int_{-\infty}^{\infty} |\xi|^{\alpha+1}|\mathcal{F}(u)(\xi)|d\xi \le C \int_{-\infty}^{\infty} |\xi|^{\alpha+1-4}d\xi < \infty.$$

Under which circunstances do we have $|\mathcal{F}(u)(\xi)| \le C|\xi|^{-4}$?

Let us assume that all derivatives exist until the fourth order and that $u^{(4)} \in L_1(\mathbb{R})$. The Fourier transform of the fourth order derivative $\mathcal{F}(u^{(4)})(\xi)$ exists and it is given by

$$\mathcal{F}(u^{(4)})(\xi) = (-i\xi)^4 \mathcal{F}(u)(\xi).$$

Therefore,

$$|\mathcal{F}(u)(\xi)| = (\xi)^{-4}|\mathcal{F}(u^{(4)})(\xi)|.$$

We also have that

$$|\mathcal{F}(u^{(4)})(\xi)| \le \frac{1}{2\pi} \int_{-\infty}^{\infty} |u^{(4)}(x)|dx \le C\|u^{(4)}\|_{L_1(\mathbb{R})}.$$

Therefore,

$$|\mathcal{F}(u)(\xi)| \le C|\xi|^{-4}.$$

4.2 The truncation error is straightforward by taking in consideration the truncation error of the approximation of the time derivative and the truncation error of the approximation of the spacial derivative. The error of the spatial derivative is given in Theorem 4.1.

4.3 (a) Write the coefficients as

$$g_{k+1}^{\alpha} = (-1)^{k+1}\binom{\alpha}{k+1} = \frac{(-1)^{k+1}\Gamma(\alpha+1)}{\Gamma(k+2)\Gamma(\alpha-k-1+1)},$$

$$g_k^{\alpha} = (-1)^k\binom{\alpha}{k} = \frac{(-1)^k\Gamma(\alpha+1)}{\Gamma(k+1)\Gamma(\alpha-k+1)}.$$

Then it is easy to conclude the recurrence formula.

(b) Note that

$$(1 + z)^\alpha = \sum_{k=0}^{\infty} \binom{\alpha}{k} z^k$$

and substitute $z = 1$.

4.4 Just proceed similarly to what we have done for the implicit Euler method in Theorem 2, that is, use the von Neumann analysis and evaluate the amplification factor for $\theta = \pi$ to see if this is larger than 1.

4.5 Replicate the derivation of the numerical method presented in the text, but now with the equation with the source term. By doing this, additional terms with the derivative of the source term will appear.

4.6 Just follow the same steps as in Section 4.3.1, but now considering

$$D_-^\alpha u(x_j) = \frac{1}{\Gamma(2 - \alpha)} \frac{d^2}{dx^2} \int_{x_j}^{\infty} u(\xi)(\xi - x_j)^{1-\alpha} d\xi.$$

Then do the same second order approximation for the second order derivative outside the integral. Afterwords we must approximate

$$I_-^{2-\alpha} u(x_j) = \frac{1}{\Gamma(2 - \alpha)} \int_{x_j}^{\infty} u(\xi)(\xi - x_j)^{1-\alpha} d\xi.$$

We proceed similarly by approximate the function inside the integral by a linear spline. In the end we will arrive at the approximation

$$D_-^\alpha u(x_j) \approx \frac{(\Delta x)^{-\alpha}}{\Gamma(4 - \alpha)} \sum_{m=-1}^{\infty} q_m^\alpha u(x_{j+m}),$$

where the coefficients q_m^α are given by (4.50).

4.7 We have

$$D_+^\alpha u(x) = \frac{1}{\Gamma(1 - \alpha)} \frac{d}{dx} \int_{-\infty}^{x} u(\tau)(x - \tau)^{-\alpha} d\tau.$$

By a change of variable

$$D_+^\alpha u(x) = \frac{1}{\Gamma(1 - \alpha)} \frac{d}{dx} \int_0^{\infty} u(x - \tau)\tau^{-\alpha} d\tau$$

$$= \frac{1}{\Gamma(1 - \alpha)} \int_0^{\infty} u'(x - \tau)\tau^{-\alpha} d\tau.$$

This integral exists since the function behaves as $|x|^{\alpha-1-\epsilon}$ in the infinity and therefore we have

$$\lim_{a \to \infty} a^{-\alpha} a^{\alpha - \epsilon} = 0.$$

4.8 Let $\beta \in (0, 1)$ and $k \geq 1$. We have $k - 1 < k - \beta < k$. Then

$$D_+^{k-\beta} u(x) = \frac{1}{\Gamma(\beta)} \frac{d^k}{dx^k} \int_{-\infty}^{x} u(\tau)(x - \tau)^{-1+\beta} d\tau.$$

Therefore,

$$D^m[D_+^{k-\beta} u(x)] = \frac{1}{\Gamma(\beta)} \frac{d^{m+k}}{dx^{m+k}} \int_{-\infty}^{x} u(\tau)(x - \tau)^{-1+\beta} d\tau = D_+^{m+k-\beta} u(x).$$

Let $\alpha := n - \beta$. We have

$$D^m[D_+^{\alpha} u(x)] = D_+^{m+\alpha} u(x).$$

4.9 Since $q_{-1}^{\alpha} + q_1^{\alpha} \geq 0$ and $q_m^{\alpha} \geq 0$, for all $m \geq 2$, then

$$\sum_{m=-1}^{\infty} q_m^{\alpha} \cos(m\phi) \leq (q_{-1}^{\alpha} + q_1^{\alpha}) + q_0^{\alpha} + \sum_{m=2}^{\infty} q_m^{\alpha} = 0.$$

4.10 (a) Similar to what has been done for the case we have a first order approximation that lead to (4.16).

(b) Similar to Theorem 4.4 regarding the von Neumann analysis, when the coefficients of the approximation of the fractional derivative are g_k^{α} instead of q_m^{α}.

Chapter 5
Fractional Differential Equations in Bounded Domains: Absorbing Boundaries

In this chapter, our focus shifts to the fractional diffusion equation with a left absorbing boundary condition. Our primary objective is to study how the presence of this boundary condition can change the characteristics of the unbounded problem that we have explored in the preceding chapter. To solve this fractional differential problem, we employ numerical methods similar to those discussed earlier and it becomes evident that the attributes of these numerical methods undergo a transformation, specifically in terms of their consistency and convergence, due to the non-local nature of the fractional operator. We establish that, in the presence of boundaries, the order of accuracy of the truncation error of the numerical methods may decrease, and in certain instances, we encounter inconsistency. This phenomenon depends not solely on the regularity of the solution but rather on the values of its derivatives at the boundary. We apply the matrix analysis and the energy method to investigate stability, and we show that, in most cases, the presence of boundaries does not significantly impact stability. Moreover, it is worth noting that the rate of convergence can surpass the order of accuracy predicted by the consistency analysis. At times, we even manage to regain the expected order observed in problems defined on the real line, when applying the respective discretizations. In particular, convergence is attained in certain cases that were initially deemed inconsistent.

5.1 Formulation of the Problem with an Absorbing Boundary Condition

In recent years, great attention has been paid to fractional differential equations that describe Lévy flights. It is now well established that, in the absence of boundaries, these equations serve as valuable tools for the study of Lévy flights, as illustrated in the preceding chapters. In the presence of boundaries, the applicability of the fractional derivatives formalism still has many open questions and has sparked an extensive discussion within the literature [8, 33, 48, 99, 150]. This discussion is related to the fact that Lévy flights are stochastic processes characterized by the occurrence of extremely long jumps and these lengthy jumps introduce specific challenges when attempting to characterize the model in scenarios involving boundary conditions.

The scenario involving an absorbing boundary condition has been studied by several authors [21, 39, 41, 49, 156, 178, 179]. In the context of a process characterized by extremely long jumps, a significant ambiguity arises regarding the precise definition of the absorbing boundary condition. The key question revolves around whether a test particle should be considered absorbed when it arrives precisely at the boundary or when it crosses the boundary at any point during a non-local jump. This question has a straightforward answer when dealing with a narrow jump length distribution, as all steps are small, and the particle cannot jump across a specific point. However, the situation becomes less evident when long jumps are a part of the process. Nonetheless, up to the present moment, the prevailing consensus has been to regard jumps across the boundary as prohibited, leading to the immediate absorption of the particle. The problem we examine here adheres to this accepted approach.

The earliest paper we could find concerning Lévy flights in the context of absorbing boundary conditions, dates back to 1995 [179]. In this paper, the definition of the absorbing boundary condition, when in the presence of Lévy flights, is established by assuming that the particle initiates its trajectory close to the origin, and the boundary is effective at all times. This problem arises naturally in the context of chemical reactions controlled by enhanced diffusion. It is worth noting that in this work the authors do not relate the model with fractional differential equations and most of the discussion is conducted using the characteristic function known as the propagator in Fourier space.

One of the earliest works we discovered in which absorbing boundary conditions for Lévy flights are formulated within the framework of fractional diffusion equations is [33].

Numerous studies addressing Lévy flights in the context of absorbing boundaries have emerged, often exploring associated questions such as first passage time, nearest distance from an absorbing boundary, and the calculation of barrier crossing events [5, 47, 150, 171].

The model considered in this chapter assumes an absorbing boundary at $x = 0$. To make sure that jumps across the boundary are forbidden, the solution of the problem is set to be zero on the left semi-axis, signifying the immediate removal of the particle upon crossing the point $x = 0$ as discussed, for instance, in [33].

We start to describe how the problem formulation undergoes alterations when transitioning from an open domain to a bounded domain featuring an absorbing boundary.

Let us consider the fractional derivative, of order $\alpha \in (1, 2)$, associated to anomalous diffusion in the context of Lévy flights and defined in the real line as the inverse Fourier transform of $(-ik)^\alpha \hat{u}(k, t)$, for a fixed t, where $\hat{u}(k, t)$ is the Fourier transform of $u(\cdot, t)$. As we have seen this leads to the definition

$$\frac{\partial^\alpha u}{\partial x^\alpha}(x, t) = \frac{1}{\Gamma(2 - \alpha)} \frac{\partial^2}{\partial x^2} \int_{-\infty}^{x} u(\xi, t)(x - \xi)^{1-\alpha} d\xi, \tag{5.1}$$

where $\Gamma(\cdot)$ is the Gamma function. The respective initial value problem is given by

$$\frac{\partial u}{\partial t}(x, t) = D \frac{\partial^\alpha u}{\partial x^\alpha}(x, t) + f(x, t), \quad x \in \mathbb{R},\ t > 0, \tag{5.2}$$

where D is the diffusion coefficient and f is a source term.

The presence of a boundary, on the left hand side of the domain, will modify the nonlocal spatial operator (5.1). This means that the boundary condition cannot be uncoupled from the fractional partial differential equation as when the order of the space derivative is an integer. Hence, by assuming that

$$u(x, t) = 0, \quad \text{for} \quad x \le 0,$$

we obtain,

$$\frac{\partial^\alpha u}{\partial x^\alpha}(x, t) = \frac{1}{\Gamma(2 - \alpha)} \frac{\partial^2}{\partial x^2} \int_{-\infty}^{x} u(\xi, t)(x - \xi)^{1-\alpha} d\xi$$

$$= \frac{1}{\Gamma(2 - \alpha)} \frac{\partial^2}{\partial x^2} \int_{0}^{x} u(\xi, t)(x - \xi)^{1-\alpha} d\xi.$$

We define the left Riemann-Liouville fractional derivative of order $\alpha \in (1, 2)$ and starting at $a = 0$ as

$$\frac{\partial_{abs}^\alpha u}{\partial x^\alpha}(x, t) = \frac{1}{\Gamma(2 - \alpha)} \frac{\partial^2}{\partial x^2} \int_{0}^{x} u(\xi, t)(x - \xi)^{1-\alpha} d\xi. \tag{5.3}$$

Formally we want to solve the problem, for $\alpha \in (1, 2)$,

$$\frac{\partial u}{\partial t}(x, t) = D \frac{\partial_{abs}^\alpha u}{\partial x^\alpha}(x, t) + f(x, t),\ x > 0,\ t > 0 \tag{5.4}$$

with an initial condition

$$u(x, 0) = u_0(x), \quad \text{for all} \quad x > 0 \tag{5.5}$$

and the boundary condition

$$u(0, t) = 0, \quad \text{for all} \quad t > 0. \tag{5.6}$$

On the right hand side of the domain we can assume

$$\lim_{x \to \infty} u(x, t) = g(t), \quad \text{for all} \quad t > 0. \tag{5.7}$$

The boundary becomes $u(0, t) = 0$, for all $t > 0$, as written in (5.6), since the condition $u(x, t) = 0$, when $x < 0$, is contained in the reformulation of the fractional operator (5.3). We can also consider a problem defined in a bounded domain $\Omega = (0, b)$ and on the right hand side of the domain we can assume

$$u(b, t) = g(t), \quad \text{for all} \quad t > 0. \tag{5.8}$$

The problem defined by Eqs. (5.4)–(5.7) entails the use of the left Riemann-Liouville fractional derivative. Consequently, it is important to note that the behavior at the right-hand side of the boundary will not significantly impact the core discussion that we are about to present.

Numerical techniques for solving space fractional partial differential equations have been documented in various references, such as, [39, 41, 45, 68, 74, 82, 88, 171, 173]. While many of these works carry out an analysis of the stability of the numerical methods, a systematic analysis on the consistency and convergence of the numerical methods, that takes into account the influence of the bounded domain, remains notably absent in the majority of these studies. In most cases, the analyses rely on consistency results established for the real line. However, as we will show, these results are significantly affected by the presence of boundaries. The primary objective of this chapter is to elucidate the distinctions of having a numerical method for an open domain or for a bounded domain.

The well-posedness of fractional differential problems with boundaries, particularly with regard to the regularity and existence of solutions, whether in the classical or weak sense, has been thoroughly studied in the existing literature [66, 87, 115, 154, 175]. Within this extensive body of research we can find the regularity of solutions in weighted Sobolev spaces and Hilbert spaces discussed in [66, 87, 175]. Furthermore, in [115], the regularity up to the boundary of solutions to the Dirichlet problem for the fractional Laplacian is studied. By the fractional Laplacian, we mean the operator $(-\Delta)^{\sigma} u$, defined via the Fourier transform $\mathscr{F}[(-\Delta)^{\sigma} u](\xi) = |\xi|^{2\sigma} \mathscr{F}(u)(\xi)$ (see Exercise 2.8 in Chap. 2). This work demonstrates that if u is a solution of the problem defined by the fractional Laplacian that is equal to a source term $g \in L_{\infty}$ in an open set and outside this set is zero, then $u \in C^{\alpha/2}(\mathbb{R})$, where $C^{\alpha/2}(\mathbb{R})$ represents the set of Hölder continuous functions. The existence of the classical solution is tackled in [154]. This work not only discusses the existence of solutions but also characterizes them by employing solutions to second-order elliptic differential equations.

Discussions on the regularity of fractional diffusion problems in variational form with homogeneous boundary conditions can be found in [52–54]. In [52, 54], the fractional operator is defined neither in terms of the Riemann-Liouville fractional derivative nor the Caputo fractional derivative, but rather through an operator involving one derivative outside the integral and another inside the integral. Further results on the regularity of weak solutions of problems with homogeneous boundary conditions are provided in [72, 73], where [73] also addresses the regularity of strong solutions.

The work more closely related to the topics under discussion in this chapter, specifically pertaining to fractional operators involving both left and right Riemann-Liouville fractional derivatives, can be found in the references [6, 7]. In these references, the well-posedness of the associated Cauchy problems is rigorously demonstrated, particularly when the initial condition lies within the spaces $C_0(\Omega)$ or $L_1(\Omega)$. It is worth noting that the space $C_0(\Omega)$ is the space of continuous functions that converge to zero at the boundary and if Ω is compact, then $C_0(\Omega)$ is the space of continuous functions on Ω. Essentially $C_0(\Omega)$ is the Banach space of continuous functions vanishing at infinity and endowed with the supremum norm. The pivotal outcome of this research is the establishment of unique solutions and these solutions possess the property of belonging to the space $C_0(\Omega)$.

As explored in the second chapter, it becomes evident that for problems defined on the real line, the solutions often exhibit a high degree of smoothness. This characteristic arises from their representation as the convolution of the initial solution with the fundamental solution of the fractional differential equation. However, if we change into scenarios involving boundaries, a significant transformation can occur in the solutions. The presence of boundaries can generate solutions that are continuous but have unbounded derivatives near the boundary.

In order to gain insights into the behavior of fractional differential problems with boundaries, even in cases where exact solutions remain beyond reach, it is instructive to turn our attention to a scenario without the time derivative.

For $\alpha \in (1, 2)$, consider the equation

$$\frac{\partial^\alpha_{abs} u}{\partial x^\alpha}(x, t) = 0.$$

Two obvious solutions emerge for this equation: $x^{\alpha-2}$ and $x^{\alpha-1}$. If we impose the condition $u(0, t) = 0$, this will force the solution to behave according to $x^{\alpha-1}$. As for the other solution $x^{\alpha-2}$, we can observe that its fractional derivative of order $\alpha - 1$ will be zero (see Exercise 5.1). This particular solution is closely related with a distinctive type of boundary condition, one that we will discuss in the following chapter, referred to as the fractional Neumann boundary condition.

The behaviour of solutions for the fractional diffusion problem with absorbing boundary conditions will depend, as usual, on the properties of the source term and of the initial condition. In the case the solutions behave as $x^{\alpha-1}$, they are expected to be continuous, specifically belonging to $C[0, 1]$, yet characterized by $C^\infty(0, 1]$. This means the derivatives of these solutions tend to become unbounded in the vicinity of the boundary.

Consider the general problem, that consists on the fractional advection diffusion equation, defined in $x > 0$, with an absorbing boundary at $x = 0$. The influence of this absorbing boundary condition is primarily limited to the left Riemann-Liouville fractional derivative. The equation for this scenario is given by

$$\frac{\partial u}{\partial t}(x,t) + V\frac{\partial u}{\partial x}(x,t) = D\left(\frac{1+\beta}{2}\frac{\partial^\alpha_{abs}u}{\partial x^\alpha}(x,t) + \frac{1-\beta}{2}\frac{\partial^\alpha u}{\partial(-x)^\alpha}(x,t)\right), \qquad (5.9)$$

with $\alpha \in (1,2]$ and $\beta \in [-1,1]$.

In the process of simulating the computational domain, it becomes necessary to establish the computational boundary on the right-hand side. This boundary condition is typically set to zero, aligning with the underlying assumption that as x tends towards infinity, the solution u converges to zero.

5.2　Accuracy of the Fractional Derivatives Approximations

In this section, we discuss how accurate are the previously examined approximations within the context of an open domain. It is important to note that a new challenge arises as a consequence of defining the left Riemann-Liouville fractional derivative with a starting point at $a = 0$.

In the context of an open domain, it is reasonable to assume that the solution approaches zero as the spatial variable extends to infinity. This assumption aligns with the origin of fractional diffusion equations, derived from the characteristic function associated with Lévy flights, as elaborated in the second chapter. Furthermore, we have also seen that continuous initial conditions yield exceptionally smooth and bounded solutions.

In the case of the well-behaved solutions in open domains, we have seen that we can derive approximations, some of which are first-order accurate, while others attain second-order accuracy. However, with the transition to a bounded domain, two fundamental questions arise. Firstly, we wonder whether the level of accuracy remains the same for solutions that are smooth enough in the closed domain. The second question is to know what happens when we allow a solution to have unbounded derivatives near the boundary. This is particularly relevant because, in various problems, solutions with unbounded derivatives are expected when dealing with bounded domains. Therefore, understanding how the approximation methods adapt to such characteristics is important.

We will see that, in general, the accuracy of fractional derivatives approximations, designed for functions across the entire real line, tends to diminish when we are in the presence of a bounded domain. Specifically, retaining the regularity of solutions within the closed domain is insufficient to preserve the same order of accuracy we observe in infinite domains. Moreover, when we permit solutions to have unbounded derivatives at the boundary, we usually need a distinct accuracy analysis.

We present a roadmap for the discussion of the consistency of any approximation to a fractional derivative. The discussion should include three distinct steps:

(a) Consistency in an open domain: The study of the consistency of the approximation in an open domain assuming we have a sufficiently enough smooth solution.
(b) Consistency in a bounded domain: The study of the consistency of the approximation within a bounded domain, with the assumption of having a sufficiently smooth solution.
(c) Consistency for solutions with unbounded derivatives at the boundary: The study of the consistency of the approximation for functions whose derivatives are unbounded near the boundary. Considering that the boundary is situated at $x = 0$, these are functions that exhibit a behavior similar to $u(x) = x^\gamma$ as x approaches zero.

In the preceding chapter, we have addressed step (a). The current chapter is dedicated to going through steps (b) and (c), providing a comprehensive exploration of the consistency of fractional derivative approximations.

Indeed the change of the domain has an impact on the accuracy of the fractional derivative approximation, effectively influencing the very essence of the approximation formula. This situation draws parallels with challenges encountered in classical differential equations, where similar issues arise. For instance, when we need to consider numerical boundary conditions [1, 64, 65, 143], or when seeking solutions with unbounded derivatives at the boundary [55, 96, 163], we encounter similar dilemmas. In the case of numerical boundaries we may end up with an approximation of lower order because we need to choose a lower order scheme to approximate the nearest points to the boundary. In the latter case, this diminished accuracy can be attributed to the inherent weak singularity exhibited by the solution or its derivatives in proximity to the boundary.

As an illustration of the first case, let us consider the transport equation that moves to the right, a problem well-posed for $t > 0$ without any boundary condition given at $x = 0$. However, the challenge arises when we employ difference approximations, particularly those employing centered difference operators for approximating the first-order derivative. At $x = 0$, these approximations break down, as no values are defined for $x < 0$. To address this issue, one approach is to use one-sided operators at the boundaries, or alternatively, to employ some form of extrapolation. For various reasons, such as stability considerations, methods that yield one-order lower accuracy than those used at inner points are often chosen (see Exercise 5.2). This preference prompts the question of whether, despite this reduction in accuracy at the boundaries, the overall accuracy of the solution of the difference approximation can be maintained at an higher order. Gustafsson [64, 65] studied this matter several years ago, and noticed that this can be the case for certain examples. Specifically, [64, 65] show that we can lower the order of the approximation near the boundary, by one or two orders, depending on whether the problem involves a first or second order derivative, respectively, and not destroy the overall accuracy of the scheme. To elucidate further, this means that when we have a lower order approximation of $m - s$ near the boundary, where m is the order of the interior scheme, we observe the following effects. For the hyperbolic problem, with the highest spatial derivative one, the full scheme will have an order of $m - s + 1$ and for the parabolic case, with the highest spatial derivative two, the full scheme will have an order of $m - s + 2$. See Exercise 5.2 regarding the classical problems.

Extending the previous idea to our specific case, which involves a fractional diffusion equation, characterized by the highest spatial derivative α, our expectation is that the full numerical scheme should be able to reach an order of $m - s + \alpha$. Remarkably, this expectation closely aligns with what we observe in practice.

In fact, for certain functions, we have no consistency due to the presence of the boundary. However, we will discover that the numerical method still manages to converge. Therefore, when dealing with fractional differential equations defined in bounded domains, the lost of accuracy of the discretization of the fractional derivative may not be visible through the numerical experiments on the convergence of the numerical methods. Hence, the numerical methods can behave better than predicted by the solo observation of the consistency of the fractional derivative approximation.

In the upcoming section, we discuss the two approximations we have seen to be of first order and second order respectively, if the domain is the real line.

5.2.1 Grünwald-Letnikov Fractional Derivative Approximation

In this section, we use the notation introduced in the first chapter. Given that our focus is on the discussion of the approximations for the fractional derivative, we do not need to consider functions that depend on time at this moment. Extending these results to functions that incorporate a temporal component is straightforward.

Let us narrow our discussion to the left Riemann-Liouville fractional derivative. The case for the right Riemann-Liouville fractional derivative is similar if a boundary is at the right hand side of the domain.

The left Riemann-Liouville fractional derivative on the domain $a < x < b$, for $n - 1 < \alpha < n$, is defined by

$$D_{a+}^{\alpha} u(x) = \frac{1}{\Gamma(n - \alpha)} \frac{d^n}{dx^n} \int_a^x u(\xi)(x - \xi)^{n-1-\alpha} d\xi. \tag{5.10}$$

We denote by $D_+^{\alpha} u(x)$ if $a = -\infty$.

For any $p \geq 0$, define the shifted Grünwald-Letnikov difference operator by

$$D_{a,p}^{\alpha,\Delta x} u(x) = \frac{1}{(\Delta x)^{\alpha}} \sum_{k=0}^{N} g_k^{\alpha} u(x - (k - p)\Delta x), \quad N = \left[\frac{x - a}{\Delta x} \right], \tag{5.11}$$

where

$$g_k^{\alpha} = (-1)^k \binom{\alpha}{k}.$$

We have seen that, for $\alpha \in (1, 2]$, if $u, u^{(2)}, |k|^{\alpha+1}|\hat{u}(k)| \in L_1(\mathbb{R})$ and $I_+^{2-\alpha}u \in AC^1(\mathbb{R})$, then the left Riemann-Liouville fractional derivative given by (5.10), for $a = -\infty$, satisfies

$$D_+^\alpha u(x) = D_{-\infty,p}^{\alpha,\Delta x} u(x) + \mathcal{O}(\Delta x). \tag{5.12}$$

This result is mainly presented in the works [97, 145, 148] and has been proved in Theorem 4.1, Chap. 4.

We want to examine the impact on the accuracy order of this type of approximation when employed for the left Riemann-Liouville fractional derivative starting at a finite point $x = a$.

Results for the case we have a bounded domain can be found in [136] for a sufficiently regular function or for a function that has unbounded derivatives near the boundary in [35, 137, 168]. Additional discussions on discretizations of fractional operators, related to the Grünwald-Letnikov formulation, can also be found in [51, 89, 90] and for particular cases of functions in [59, 104, 132].

In this section, our primary objective is to study the truncation error, defined as the difference between the Riemann-Liouville fractional derivative (5.10) and the Grünwald-Letnikov approximation (5.11). In other words, we discuss the behaviour of the function $\epsilon_T(\Delta x, \alpha, a, x)$ that satisfies

$$D_{a,p}^{\alpha,\Delta x} u(x) = D_{a+}^\alpha u(x) + \epsilon_T(\Delta x, \alpha, a, x).$$

Suppose that we want to apply the result (5.12), proved for a function defined in $\mathbb{R}$, to the case where the function is defined in a bounded domain $[a, b]$. One way to apply this result is to assume that

$$D_{-\infty}^\alpha u_{ext}(x) = D_{a+}^\alpha u(x), \tag{5.13}$$

where u_{ext} is an extension of the function u, defined in $\mathbb{R}$, by $u_{ext}(x) = u(x)$ for $x \in [a, b]$ and zero otherwise. However, for $\alpha \in (1, 2)$, under the previous assumptions, we must have at least

$$u_{ext}^{(r)}(a) = 0, \quad r = 0, 1. \tag{5.14}$$

We are dealing with the left fractional derivative, a non-local operator, and to compute the derivative at a certain point x, we need to take in consideration all points within the domain that are less than or equal to x. The right boundary does not play any role in this calculation. On the other hand, the left boundary determines the size of the domain that is taken into account to compute the fractional derivative at each point x.

We can now ask about what happens when conditions (5.14) are not verified and if in these cases the Grünwald-Letnikov approximation is still consistent.

With that in mind suppose that the function keeps the regularity stated in the open domain case. What we observe is that in a bounded domain, we can lose the order of accuracy one. The change from open to closed domain changes the main result independently of how smooth are the solutions. This highlights how different fractional differential problems can be from classical differential problems.

We present the result regarding the consistency of the Grünwald-Letnikov approximation, for a general $n - 1 < \alpha < n$. The proof is not included here due to its extension and technical complexity, but it can be found with great detail in [136].

Theorem 5.1 *Let $n - 1 < \alpha < n$, $u \in AC^{n+1}([a, b])$, $u^{(n+1)} \in L_\infty(a, b)$ and*

$$D_{a+}^\alpha u(x) = D_{a,p}^{\alpha, \Delta x} u(x) + \epsilon_T(\Delta x, \alpha, a, x).$$

Then, for Δx small enough, we have

$$|\epsilon_T(\Delta x, \alpha, a, x)| \le \sum_{k=0}^{n-1} C_k \Delta x (x - a)^{k-1-\alpha} |u^{(k)}(a)|$$

$$+ \sum_{k=1}^{n-1} \overline{C}_k \Delta x (x - a)^{k-\alpha} |u^{(k+1)}(a)|$$

$$+ (\tilde{C}_1 (\Delta x)^{n-\alpha} + \tilde{C}_2 \Delta x) ||u^{(n)}||_\infty + (\tilde{C}_3 \Delta x (x - a)^{n-\alpha}$$

$$+ \tilde{C}_4 (\Delta x)^{1+n-\alpha}) ||u^{(n+1)}||_\infty$$

$$+ \mathcal{O}((\Delta x)^{n+1-\alpha}),$$

where C_k, $k = 0, 1, \ldots n - 1$, $\overline{C}_k$, $k = 1, \ldots n - 1$ and $\tilde{C}_k$, $k = 1, 2, 3, 4$ are constants that do not depend on Δx.

The assertion in Theorem 5.1 tell us that for functions defined in a bounded domain, if the approximation is consistent, the order of accuracy will be in general, for $n-1 < \alpha < n$, between $n-\alpha$ and 1, depending on the properties of the functions. If $u^{(n-1)}(a) \neq 0$ we can have inconsistency. his result highlights the relevance of the left boundary point on the accuracy of the approximation.

In particular, the approximation (5.11) of the Riemann-Liouville fractional derivative (5.10), on the domain $[a, b]$, is consistent if

$$u^{(k)}(a) = 0, \quad k = 0, 1, \ldots, n - 1.$$

For $u^{(n)}(a) \neq 0$, the order of accuracy is at least $n - \alpha$.

If we allow weakly singular functions, which are continuous but may exhibit singularities in some of its derivatives at the boundary point, we can establish the following result, that has been given in [132, Proposition 3] and in Lemma 2.1 in [168, Lemma 2.1].

Theorem 5.2 *Let $u(x) = x^\gamma$, $\gamma > 0$. Then, for $\alpha > 0$,*

$$D_{0+}^\alpha u(x_j) = D_{0,p}^{\alpha,\Delta x} u(x_j) - \Delta x \left(p - \frac{\alpha}{2} \right) \frac{\Gamma(\gamma+1)}{\Gamma(\gamma-\alpha)} x_j^{\gamma-1-\alpha} + (\Delta x)^2 R^{j,\alpha,\gamma},$$

where $j \geq |q|$, q is an integer and $|R^{j,\alpha,\gamma}| \leq C x_j^{\gamma-2-\alpha}$.

Proof From (5.11), we write

$$D_{a,p}^{\alpha,\Delta x} u(x_j) = \frac{1}{(\Delta x)^\alpha} \sum_{k=0}^{j} g_k^\alpha u(x_j - (k-p)\Delta x).$$

Let $p = 0$. Hence

$$(\Delta x)^{\alpha-\gamma} D_{a,0}^{\alpha,\Delta x} u(x_j) = \sum_{k=0}^{j} g_k^\alpha (j-k)^\gamma = \sum_{k=0}^{j} g_{j-k}^\alpha k^\gamma.$$

We have that (see Exercise 5.3)

$$\sum_{k=0}^{j} g_{j-k}^\alpha k^\gamma = \frac{\Gamma(\gamma+1)}{\Gamma(\gamma+1-\alpha)} j^{\gamma-\alpha} - \frac{\alpha}{2} \frac{\Gamma(\gamma+1)}{\Gamma(\gamma-\alpha)} j^{\gamma-\alpha-1} + \mathcal{O}(j^{\gamma-\alpha-2}).$$

$$(5.15)$$

This yields the result we are looking for when $p = 0$.

For $p \neq 0$, we have

$$(\Delta x)^{\alpha-\gamma} D_{a,p}^{\alpha,\Delta x} u(x_j) = \sum_{k=0}^{j+p} g_k^\alpha (j-k+p)^\gamma = \sum_{k=0}^{j+p} g_{j+p-k}^\alpha k^\gamma.$$

From the equality (5.15), it follows

$$\sum_{k=0}^{j+p} g_{j+p-k}^\alpha k^\gamma = \frac{\Gamma(\gamma+1)}{\Gamma(\gamma+1-\alpha)} (j+p)^{\gamma-\alpha} - \frac{\alpha}{2} \frac{\Gamma(\gamma+1)}{\Gamma(\gamma-\alpha)} (j+p)^{\gamma-\alpha-1}$$
$$+ \mathcal{O}((j+p)^{\gamma-\alpha-2}).$$

By the fact that (see Exercise 5.4)

$$\left(1 + \frac{p}{j} \right)^\gamma = 1 + \gamma \frac{p}{j} + \mathcal{O}(j^{-2}), \quad j \geq |p|,$$

we have

$$\sum_{k=0}^{j+p} g_{j+p-k}^{\alpha} k^{\gamma} = \frac{\Gamma(\gamma+1)}{\Gamma(\gamma+1-\alpha)} j^{\gamma-\alpha} \left(1 + (\gamma-\alpha)\frac{p}{j} + \mathcal{O}(j^{-2})\right)$$
$$- \frac{\alpha}{2} \frac{\Gamma(\gamma+1)}{\Gamma(\gamma-\alpha)} j^{\gamma-\alpha-1} \left(1 + (\gamma-\alpha-1)\frac{p}{j} + \mathcal{O}(j^{-2})\right)$$
$$+ \mathcal{O}((j+p)^{\gamma-\alpha-2}).$$

Since $\Gamma(\gamma+1-\alpha) = (\gamma-\alpha)\Gamma(\gamma-\alpha)$, then

$$\sum_{k=0}^{j+p} g_{j+p-k}^{\alpha} k^{\gamma} = \frac{\Gamma(\gamma+1)}{\Gamma(\gamma+1-\alpha)} j^{\gamma-\alpha} + \frac{\Gamma(\gamma+1)}{\Gamma(\gamma-\alpha)} j^{\gamma-\alpha-1} p$$
$$- \frac{\alpha}{2} \frac{\Gamma(\gamma+1)}{\Gamma(\gamma-\alpha)} j^{\gamma-\alpha-1} + \mathcal{O}((j+p)^{\gamma-\alpha-2}).$$

Therefore, we obtain

$$D_{a,p}^{\alpha,\Delta x} u(x_j) = (\Delta x)^{\gamma-\alpha} \sum_{k=0}^{j+p} g_{j+p-k}^{\alpha} k^{\gamma}$$
$$= \frac{\Gamma(\gamma+1)}{\Gamma(\gamma+1-\alpha)} (x_j)^{\gamma-\alpha} + \Delta x \left(p - \frac{\alpha}{2}\right) \frac{\Gamma(\gamma+1)}{\Gamma(\gamma-\alpha)} (x_j)^{\gamma-\alpha-1}$$
$$+ \mathcal{O}((j+p)^{\gamma-\alpha-2}).$$

$\square$

The attention given to this particular set of functions is due to the fact that the solutions of some fractional differential problems can include terms x^{γ} in its decomposition as we have mentioned before.

Consider $\alpha \in (1, 2)$ and the function $u(x) = x^{\gamma}$, $\gamma \in (2, 3)$, for which we have $u(0) = u'(0) = u''(0) = 0$. The order of accuracy according to Theorem 5.2 is $\gamma - \alpha$ and therefore is not 1. This can be easily checked by computing the local truncation error at the first node $x_1 = \Delta x$. We have

$$D_{0+}^{\alpha} u(x_1) - D_{0,p}^{\alpha,\Delta x} u(x_1) = \Delta x\, C_{\alpha,\gamma,p} (\Delta x)^{\gamma-1-\alpha} + (\Delta x)^2 R^{1,\alpha,\gamma},$$

that is

$$D_{0+}^{\alpha} u(x_1) - D_{0,p}^{\alpha,\Delta x} u(x_1) = \mathcal{O}((\Delta x)^{\gamma-\alpha}).$$

We turn to the discussion of the global truncation errors in the discrete L_∞ and L_2 norms for the typical case of having a solution behaving as x^γ near the left boundary.

Consider the discrete set of points

$$\{x_j = j\Delta x : j = 0, 1, \ldots, N\}$$

defined in $[0, b]$, with $x_0 = 0$ and $x_N = b$. For $j = 1, 2, \ldots, N$, define

$$\epsilon_T(x_j) = D_{0+}^\alpha u(x_j) - D_{0,p}^{\alpha,\Delta x} u(x_j).$$

The discrete L_∞ and L_2 norms of the vector $\epsilon_\mathbf{T} = (\epsilon_T(x_1), \epsilon_T(x_2), \ldots, \epsilon_T(x_N))$ are respectively given by

$$||\epsilon_\mathbf{T}||_\infty = \max_{1 \le j \le N} \{\epsilon_T(x_j)\}, \qquad ||\epsilon_\mathbf{T}||_2^2 = \sum_{j=1}^N \Delta x (\epsilon_T(x_j))^2.$$

We present results regarding these norms for the local truncation error. First, we state the result in the maximum norm.

Theorem 5.3 *Let* $n - 1 < \alpha < n$, $u \in AC([0, b]) \cap C^{n+1}(0, b]$ *and u near the left boundary behaves as* x^γ, $\gamma > 0$, *that is,*

$$\sup_{x \in (0,b)} x^{n-\gamma} |u^{(n)}(x)| < \infty.$$

Then, the truncation error satisfies $||\epsilon_\mathbf{T}||_\infty = \mathcal{O}((\Delta x)^q)$, $q = \min\{1, \gamma - \alpha\}$.

Proof By Theorem 5.2, the leading term of the truncation error is given by $C(\Delta x)x_j^{\gamma-1-\alpha}$, for some positive constant C. It is easy to see that if $\gamma - \alpha \ge 1$ we have order of accuracy 1. On the other hand if $\gamma - \alpha < 1$, from the result of Theorem 5.2, we obtain

$$|\epsilon_T(x_j)| \le C_1 j^{\gamma-1-\alpha}(\Delta x)^{\gamma-\alpha} + C_2 j^{\gamma-2-\alpha}(\Delta x)^{\gamma-\alpha},$$

where C_1, C_2 are positive constants. For $0 \le \gamma < 1 + \alpha$, we have $j^{\gamma-1-\alpha} \le 1$ and therefore $|\epsilon_T(x_j)| \le C(\Delta x)^{\gamma-\alpha}$. $\qquad\square$

We now prove the result for the truncation error in the discrete L_2 norm.

Theorem 5.4 *Let* $n - 1 < \alpha < n$, $u \in AC([0, b]) \cap C^{n+1}(0, b]$ *and u near the boundary behaves as* x^γ, $\gamma > 0$, *that is,*

$$\sup_{x \in (0,b)} x^{n-\gamma} |u^{(n)}(x)| < \infty.$$

Then, the truncation error satisfies

$$||\epsilon_{\mathbf{T}}||_2 = \begin{cases} \mathcal{O}(\Delta x), & for \quad \gamma - \alpha \geq 1, \\ \mathcal{O}((\Delta x)^{1-\epsilon/2}), & for \quad 1/2 \leq \gamma - \alpha < 1 \ and \ a \ small \ \epsilon > 0, \\ \mathcal{O}((\Delta x)^{\gamma-\alpha+1/2}), & for \quad \gamma - \alpha < 1/2. \end{cases}$$

Proof The leading term of the truncation error is given by $\epsilon_T^{\text{lead}}(x_j) :=$ $C(\Delta x)x_j^{\gamma-1-\alpha}$, for some positive constant C. If $\gamma - \alpha \geq 1$ the order of accuracy is 1. If $\gamma - \alpha < 1$, for the leading term, we have

$$\sum_{j=1}^{N} \Delta x(\epsilon_T^{\text{lead}}(x_j))^2 \leq C\Delta x \sum_{j=1}^{N}(\Delta x)^2 x_j^{2\gamma-2(1+\alpha)}$$

$$= C(\Delta x)^{2(\gamma-\alpha)}\Delta x \sum_{j=1}^{N} j^{2(\gamma-\alpha)-2}.$$

If $\gamma - \alpha < 1/2$, then

$$||\epsilon_{\mathbf{T}}||_2^2 \leq C(\Delta x)^{2(\gamma-\alpha)+1}\xi(2 - 2(\gamma - \alpha)),$$

where $\xi(s)$ is the Riemann zeta function, that converges when $s > 1$. Therefore

$$||\epsilon_{\mathbf{T}}||_2 \leq C(\Delta x)^{(\gamma-\alpha)+1/2}.$$

On the other hand, for $1/2 \leq \gamma - \alpha < 1$, we can rewrite, for a small $\epsilon > 0$,

$$\sum_{j=1}^{N} \Delta x(\epsilon_T^{\text{lead}}(x_j))^2 \leq C(\Delta x)^{2-\epsilon} x_N^{2(\gamma-\alpha)-1+\epsilon}\xi(1 + \epsilon),$$

where $\xi(s)$ is the Riemann zeta function. Hence

$$||\epsilon_{\mathbf{T}}||_2 \leq C_\epsilon(\Delta x)^{1-\epsilon/2},$$

for a constant C_ϵ that do not depend on Δx. $\qquad\square$

5.2.2   Linear Spline Fractional Derivative Approximation

In Chap. 4, we have introduced a second-order approximation for the Riemann-Liouville fractional derivative. In this chapter, we will adapt the previously derived approximation to the case when we have an absorbing boundary at $x = a$. For clarity purposes, in the next results we continue to omit the dependence on t of the function u and only consider a function $u(x)$. The extension to partial derivatives is trivial.

According to the notations and definitions presented in the previous chapters, we define the left Riemann-Liouville fractional derivative of order $n - 1 < \alpha < n$ by

$$D_{a+}^{\alpha} u(x) = \frac{1}{\Gamma(n - \alpha)} \frac{d^n}{dx^n} \int_a^x u(\xi)(x - \xi)^{n-1-\alpha} d\xi \tag{5.16}$$

and the left fractional integral as

$$I_{a+}^{n-\alpha} u(x) = \frac{1}{\Gamma(n - \alpha)} \int_a^x u(\xi)(x - \xi)^{n-1-\alpha} d\xi. \tag{5.17}$$

The discrete domain of $[x_0, x_N]$ is the set of points

$$x_k = x_{k-1} + k\Delta x, \quad k = 1, \ldots N,$$

where Δx is the space step.

The linear spline that interpolates the values $\{u(x_k), \ k \leq j\}$ in the interval $[x_0, x_j]$, $j = 1, \ldots, N$, is of the form

$$s^j(\tau) = \sum_{k=0}^{j} u(x_k) s_k(\tau),$$

where for $k = 1, \ldots, j - 1$, the functions s_k are the hat functions. For $k = 0$, the function s_0 is only defined in $[x_0, x_1]$ and for $k = j$ the function s_j is only defined in the interval $[x_{j-1}, x_j]$. At x_j, for $j = 1, \ldots, N$, we have the approximation of (5.17),

$$\tilde{I}_{a+}^{2-\alpha} u(x_j) = \frac{1}{\Gamma(2 - \alpha)} \sum_{k=0}^{j} u(x_k) \int_a^{x_j} (x_j - \xi)^{1-\alpha} s_k(\xi) d\xi$$

$$= \frac{(\Delta x)^{2-\alpha}}{\Gamma(4 - \alpha)} \sum_{k=0}^{j} u(x_k) a_{j,k}^{\alpha}, \tag{5.18}$$

where

$$a_{j,k}^{\alpha} = \begin{cases} (j - 1)^{3-\alpha} - j^{2-\alpha}(j - 3 + \alpha), & k = 0, \\ (j - k - 1)^{3-\alpha} - 2(j - k)^{3-\alpha} + (j - k + 1)^{3-\alpha}, & 1 \leq k \leq j - 1, \\ 1, & k = j. \end{cases}$$

Compare (5.18) with the approximation in the third section of Chap. 4.

We proceed as in the third section of Chap. 4, that is, to approximate the left Riemann-Liouville fractional derivative, the second order derivative outside the integral is approximated using the central second order difference, as in [137, 140]. Taking in consideration (5.18) and that $u(x_0) = 0$, we have

$$\frac{d^2}{dx^2} I_{a+}^{2-\alpha} u(x_j) \approx \frac{I_{a+}^{2-\alpha} u(x_{j+1}) - 2 I_{a+}^{2-\alpha} u(x_j) + I_{a+}^{2-\alpha} u(x_{j-1})}{(\Delta x)^2}$$

$$\approx \frac{1}{(\Delta x)^\alpha \Gamma(4 - \alpha)} \sum_{k=1}^{j+1} q_{j,k}^\alpha u(x_k),$$

where

$$q_{j,k}^\alpha = \begin{cases} a_{j+1,j+1}^\alpha, & k = j+1 \\ -2a_{j,j}^\alpha + a_{j+1,j}^\alpha, & k = j \\ a_{j-1,k}^\alpha - 2a_{j,k}^\alpha + a_{j+1,k}^\alpha, & 1 \le k \le j-1. \end{cases}$$

In order to simplify the writing of the previous approximation, define

$$a_0^\alpha = 1, \quad a_m^\alpha = (m+1)^{3-\alpha} - 2m^{3-\alpha} + (m-1)^{3-\alpha}, \ 1 \le m \le j-1 \qquad (5.19)$$

and

$$q_{-1}^\alpha = a_0^\alpha, \quad q_0^\alpha = a_1^\alpha - 2a_0^\alpha, \quad q_m^\alpha = a_{m+1}^\alpha - 2a_m^\alpha + a_{m-1}^\alpha, \ 1 \le m \le j-1. \qquad (5.20)$$

Then, the left Riemann-Liouville fractional derivative, for $j = 1, \ldots, N-1$, can be approximated by

$$D_{a,spl}^{\alpha, \Delta x} u(x_j) := \frac{1}{(\Delta x)^\alpha \Gamma(4 - \alpha)} \sum_{m=-1}^{j} q_m^\alpha u(x_{j-m}).$$

The notation $D_{a,spl}^{\alpha, \Delta x} u$ uses spl to refer to the linear spline approximation. If we compare this formula with the formula obtained for the real line in the third section of Chap. 4 we see that the coefficients are the same. What has changed is the sum that is now a finite sum.

We proceed to compute an upper bound for the local truncation error of the approximation of the left Riemann-Liouville fractional derivative when defined in the domain $[a, b]$. The formulation of this upper bound is of great importance, as it will subsequently serve as a crucial tool for demonstrating how inconsistency can lead to convergence in certain scenarios.

We start to derive an upper bound for the local truncation error, when $\alpha \in (1, 2)$, that is, for the quantity

$$\epsilon_T(\Delta x, \alpha, a, x_j) = D_{a+}^{\alpha} u(x_j) - D_{a,spl}^{\alpha,\Delta x} u(x_j). \tag{5.21}$$

Before proving the main result, in Theorem 5.5, about the local truncation error, and in order to enhance clarity, we first establish several auxiliary results that will be used within its proof.

Lemma 5.1 *For $j = 2, \ldots, N - 1$ define*

$$\epsilon_{2,2}(x_j) = \frac{1}{(\Delta x)^2} \frac{1}{\Gamma(4 - \alpha)} \sum_{p=0}^{2} (-1)^p \binom{2}{p}$$

$$\times \left(\sum_{k=1}^{j-1+p} \int_{x_{k-1}}^{x_k} l_{k,2}(\xi) u^{(2)}(\xi)(x_{j-1+p} - \xi)^{1-\alpha} d\xi \right),$$

with $|l_{k,2}(\xi)| \leq C(\Delta x)^2$, where C is a positive constant that do not depend on k and Δx and $l_{k\pm 1,2}(\xi + \Delta x) = l_{k,2}(\xi)$. Then we have the local bound

$$|\epsilon_{2,2}(x_j)| \leq \frac{C(\Delta x)^2}{(2 - \alpha)\Gamma(4 - \alpha)}(x_j - x_0)^{2-\alpha} \max_{x \in (x_2, x_j)} |u^{(4)}(x)|$$

$$+ \frac{C 2^{\alpha-1}(\Delta x)^2}{\Gamma(4 - \alpha)}(x_j - x_0)^{1-\alpha} \max_{x \in (x_0, x_1)} |u^{(3)}(x)|$$

$$+ \frac{C(\Delta x)^2}{\Gamma(4 - \alpha)}(x_j - x_0)^{-\alpha} \max_{x \in (x_0, x_1)} |u^{(2)}(x)|.$$

Proof By a change of variables, we obtain

$$\epsilon_{2,2}(x_j) = \frac{1}{(\Delta x)^2} \frac{1}{\Gamma(4 - \alpha)} \sum_{p=0}^{2} (-1)^p \binom{2}{p} S_p, \tag{5.22}$$

with

$$S_p = \sum_{k=1}^{j-1+p} \int_{x_{k-1}-(p-1)\Delta x}^{x_k-(p-1)\Delta x} l_{k,2}(\xi + (p-1)\Delta x) u^{(2)}(\xi + (p-1)\Delta x)$$

$$\times (x_j - \xi)^{1-\alpha} d\xi$$

$$= \sum_{k=2-p}^{j} \int_{x_{k-1}}^{x_k} l_{k+(p-1),2}(\xi + (p-1)\Delta x) u^{(2)}(\xi + (p-1)\Delta x)(x_j - \xi)^{1-\alpha} d\xi.$$

Since $l_{k\pm1,2}(\xi \pm \Delta x) = l_{k,2}(\xi)$, we have

$$\epsilon_{2,2}(x_j) = \frac{1}{(\Delta x)^2} \frac{1}{\Gamma(4-\alpha)}$$

$$\times \sum_{k=2}^{j} \int_{x_{k-1}}^{x_k} l_{k,2}(\xi) \left(u^{(2)}(\xi+\Delta x) - 2u^{(2)}(\xi) + u^{(2)}(\xi-\Delta x) \right) (x_j - \xi)^{1-\alpha} d\xi$$

$$-\frac{2}{(\Delta x)^2} \frac{1}{\Gamma(4-\alpha)} \int_{x_0}^{x_1} l_{1,2}(\xi) u^{(2)}(\xi)(x_j - \xi)^{1-\alpha} d\xi$$

$$+\frac{1}{(\Delta x)^2} \frac{1}{\Gamma(4-\alpha)} \sum_{k=0}^{1} \int_{x_{k-1}}^{x_k} l_{k,2}(\xi) u^{(2)}(\xi + \Delta x)(x_j - \xi)^{1-\alpha} d\xi. \tag{5.23}$$

We can write

$$\epsilon_{2,2}(x_j) = \frac{1}{(\Delta x)^2} \frac{1}{\Gamma(4-\alpha)} \left(A_j + A_1 \right),$$

for

$$A_j = \sum_{k=2}^{j} \int_{x_{k-1}}^{x_k} l_{k,2}(\xi) \left(u^{(2)}(\xi + \Delta x) - 2u^{(2)}(\xi) + u^{(2)}(\xi - \Delta x) \right)$$

$$\times (x_j - \xi)^{1-\alpha} d\xi,$$

$$A_1 = \int_{x_0}^{x_1} l_{1,2}(\xi) \left(u^{(2)}(\xi + \Delta x) - u^{(2)}(\xi) \right) (x_j - \xi)^{1-\alpha} d\xi$$

$$+\int_{x_{-1}}^{x_0} l_{0,2}(\xi) u^{(2)}(\xi + \Delta x)(x_j - \xi)^{1-\alpha} d\xi - \int_{x_0}^{x_1} l_{1,2}(\xi) u^{(2)}(\xi)(x_j - \xi)^{1-\alpha} d\xi.$$

Let us start to bound A_1. By a change of variables, we have that

$$\int_{x_{-1}}^{x_0} l_{0,2}(\xi) u^{(2)}(\xi + \Delta x)(x_j - \xi)^{1-\alpha} d\xi = \int_{x_0}^{x_1} l_{1,2}(\xi) u^{(2)}(\xi)(x_{j+1} - \xi)^{1-\alpha} d\xi$$

and consequently

$$A_1 = \int_{x_0}^{x_1} l_{1,2}(\xi) \left(u^{(2)}(\xi + \Delta x) - 2u^{(2)}(\xi) \right) (x_j - \xi)^{1-\alpha} d\xi$$

$$+\int_{x_0}^{x_1} l_{1,2}(\xi) u^{(2)}(\xi)(x_{j+1} - \xi)^{1-\alpha} d\xi. \tag{5.24}$$

There exists a $\xi_1 \in (x_0, x_1)$ such that

$$A_1 = - \int_{x_0}^{x_1} l_{1,2}(\xi) u^{(2)}(\xi)(x_j - \xi)^{1-\alpha} d\xi$$

$$+ \Delta x \int_{x_0}^{x_1} l_{1,2}(\xi) u^{(3)}(\xi_1)(x_j - \xi)^{1-\alpha} d\xi$$

$$+ \int_{x_0}^{x_1} l_{1,2}(\xi) u^{(2)}(\xi)(x_{j+1} - \xi)^{1-\alpha} d\xi$$

$$= \int_{x_0}^{x_1} l_{1,2}(\xi) u^{(2)}(\xi) \left((x_{j+1} - \xi)^{1-\alpha} - (x_j - \xi)^{1-\alpha} \right) d\xi$$

$$+ \Delta x \int_{x_0}^{x_1} l_{1,2}(\xi) u^{(3)}(\xi_1)(x_j - \xi)^{1-\alpha} d\xi.$$

We have that (See Exercise 5.5)

$$(x_{j+1} - \xi)^{1-\alpha} - (x_j - \xi)^{1-\alpha} \leq \Delta x (x_j - x_0)^{-\alpha}.$$

Additionally, for $j \geq 2$, we have

$$\int_{x_0}^{x_1} (x_j - \xi)^{1-\alpha} d\xi \leq \Delta x (x_j - x_1)^{1-\alpha} \leq 2^{\alpha-1} \Delta x (x_j - x_0)^{1-\alpha},$$

by observing that

$$(x_j - x_1)^{1-\alpha} = (x_j - x_0)^{1-\alpha} \left(\frac{x_j - x_0}{x_j - x_1} \right)^{\alpha-1} \leq 2^{\alpha-1}(x_j - x_0)^{1-\alpha}.$$

Therefore, since $|l_{k,2}(\xi)| \leq C(\Delta x)^2$,

$$|A_1| \leq C(\Delta x)^2 \max_{x \in (x_0, x_1)} |u^{(2)}(x)| \int_{x_0}^{x_1} \left((x_{j+1} - \xi)^{1-\alpha} - (x_j - \xi)^{1-\alpha} \right) d\xi$$

$$+ C(\Delta x)^3 \max_{x \in (x_0, x_1)} |u^{(3)}(x)| \int_{x_0}^{x_1} (x_j - \xi)^{1-\alpha} d\xi$$

$$\leq C(\Delta x)^4 (x_j - x_0)^{-\alpha} \max_{x \in (x_0, x_1)} |u^{(2)}(x)|$$

$$+ C 2^{\alpha-1}(\Delta x)^4 (x_j - x_0)^{1-\alpha} \max_{x \in (x_0, x_1)} |u^{(3)}(x)|.$$

We can bound A_j more easily. Since $u \in AC^4([a, b])$, the fourth order derivative is integrable and there exists $\xi_k \in (x_{k-1}, x_k)$ such that

$$A_j = (\Delta x)^2 \sum_{k=2}^{j} \int_{x_{k-1}}^{x_k} l_{k,2}(\xi) u^{(4)}(\xi_k)(x_j - \xi)^{1-\alpha} d\xi.$$

Since $|l_{k,2}(\xi)| \le C(\Delta x)^2$, if $u^{(4)}$ is bounded, then, for $j = 2, \ldots, N - 1$,

$$|A_j| \le \frac{C}{2 - \alpha}(x_j - x_0)^{2-\alpha}(\Delta x)^4 \max_{x \in (x_2, x_j)} |u^{(4)}(x)|.$$

Therefore, we have, for $j \ge 2$,

$$\begin{aligned}
|\epsilon_{2,2}(x_j)| \le {} & \frac{C(\Delta x)^2}{(2-\alpha)\Gamma(4-\alpha)}(x_j - x_0)^{2-\alpha} \max_{x \in (x_2, x_j)} |u^{(4)}(x)| \\
& + \frac{C 2^{\alpha-1}(\Delta x)^2}{\Gamma(4-\alpha)}(x_j - x_0)^{1-\alpha} \max_{x \in (x_0, x_1)} |u^{(3)}(x)| \\
& + \frac{C(\Delta x)^2}{\Gamma(4-\alpha)}(x_j - x_0)^{-\alpha} \max_{x \in (x_0, x_1)} |u^{(2)}(x)|.
\end{aligned}$$

$\square$

Lemma 5.2 *For $j = 2, \ldots, N - 1$, define*

$$\begin{aligned}
\epsilon_{2,3}(x_j) = {} & \frac{1}{(\Delta x)^2}\frac{1}{\Gamma(4-\alpha)}\sum_{p=0}^{2}(-1)^p \binom{2}{p} \\
& \times \left(\sum_{k=1}^{j-1+p} \int_{x_{k-1}}^{x_k} l_{k,3}(\xi)u^{(3)}(\xi)(x_{j-1+p} - \xi)^{1-\alpha}d\xi \right),
\end{aligned}$$

with $|l_{k,3}(\xi)| \le C(\Delta x)^3$, where C is a positive constant that do not depend on k and Δx and $l_{k\pm1,3}(\xi + \Delta x) = l_{k,3}(\xi)$. Then we have the local bound

$$\begin{aligned}
|\epsilon_{2,3}(x_j)| \le {} & \frac{C(\Delta x)^2}{(2-\alpha)\Gamma(4-\alpha)}(x_j - x_0)^{2-\alpha} \max_{x \in (x_2, x_j)} |u^{(4)}(x)| \\
& + \frac{2C(\Delta x)^2}{(2-\alpha)\Gamma(4-\alpha)}(x_j - x_0)^{2-\alpha} \max_{x \in (x_0, x_1)} |u^{(4)}(x)| \\
& + \frac{C(\Delta x)^2}{\Gamma(4-\alpha)}(x_j - x_0)^{1-\alpha} \max_{x \in (x_0, x_1)} |u^{(3)}(x)|.
\end{aligned}$$

Proof We start to proceed similarly to what has been done, in the previous lemma, for $\epsilon_{2,2}(x_j)$ obtaining the equivalent equalities (5.22) and (5.23). Then, we can write

$$\epsilon_{2,3}(x_j) = \frac{1}{(\Delta x)^2}\frac{1}{\Gamma(4-\alpha)}\left(B_j + B_1\right),$$

for

$$B_j = \sum_{k=2}^{j} \int_{x_{k-1}}^{x_k} l_{k,3}(\xi) \left(u^{(3)}(\xi + \Delta x) - 2u^{(3)}(\xi) + u^{(3)}(\xi - \Delta x) \right) (x_j - \xi)^{1-\alpha} d\xi,$$

$$B_1 = \int_{x_0}^{x_1} l_{1,3}(\xi) \left(u^{(3)}(\xi + \Delta x) - 2u^{(3)}(\xi) \right) (x_j - \xi)^{1-\alpha} d\xi$$

$$+ \int_{x_{-1}}^{x_0} l_{0,3}(\xi) u^{(3)}(\xi + \Delta x)(x_j - \xi)^{1-\alpha} d\xi.$$

Since $|l_{k,3}(\xi)| \leq C(\Delta x)^3$, if $u^{(4)}$ is bounded then, for $j = 2, \ldots, N-1$,

$$|B_j| \leq \frac{C}{2-\alpha}(x_j - x_0)^{2-\alpha}(\Delta x)^4 \max_{x \in (x_2, x_j)} |u^{(4)}(x)|.$$

Additionally, for B_1, by a change of variables, we have that

$$\int_{x_{-1}}^{x_0} l_{0,3}(\xi) u^{(3)}(\xi + \Delta x)(x_j - \xi)^{1-\alpha} d\xi = \int_{x_0}^{x_1} l_{1,3}(\xi) u^{(3)}(\xi)(x_{j+1} - \xi)^{1-\alpha} d\xi$$

and

$$B_1 = \int_{x_0}^{x_1} l_{1,3}(\xi) \left(u^{(3)}(\xi + \Delta x) - u^{(3)}(\xi) \right) (x_j - \xi)^{1-\alpha} d\xi$$

$$+ \int_{x_0}^{x_1} l_{0,3}(\xi) u^{(3)}(\xi) \left((x_{j+1} - \xi)^{1-\alpha} - (x_j - \xi)^{1-\alpha} \right) d\xi.$$

Similarly to what has been done for A_1, in Lemma 5.1, and since $\xi \in [x_0, x_1]$ we also have

$$\int_{x_0}^{x_1} \left((x_{j+1} - \xi)^{1-\alpha} - (x_j - \xi)^{1-\alpha} \right) d\xi \leq \Delta x (x_j - x_0)^{1-\alpha}.$$

Therefore

$$|B_1| \leq C(\Delta x)^4 \max_{x \in (x_0, x_1)} |u^{(4)}(x)| \int_{x_0}^{x_1} (x_j - \xi)^{1-\alpha} d\xi$$

$$+ C(\Delta x)^3 \max_{x \in (x_0, x_1)} |u^{(3)}(x)| \int_{x_0}^{x_1} \left((x_{j+1} - \xi)^{1-\alpha} - (x_j - \xi)^{1-\alpha} \right) d\xi$$

$$\leq \frac{2C}{2-\alpha}(\Delta x)^4 (x_j - x_0)^{2-\alpha} \max_{x \in (x_0, x_1)} |u^{(4)}(x)|$$

$$+ C(\Delta x)^4 (x_j - x_0)^{1-\alpha} \max_{x \in (x_0, x_1)} |u^{(3)}(x)|.$$

It follows that

$$|\epsilon_{2,3}(x_j)| \leq \frac{C(\Delta x)^2}{(2-\alpha)\Gamma(4-\alpha)}(x_j - x_0)^{2-\alpha}\max_{x\in(x_2,x_j)}|u^{(4)}(x)|$$

$$+\frac{2C(\Delta x)^2}{(2-\alpha)\Gamma(4-\alpha)}(x_j - x_0)^{2-\alpha}\max_{x\in(x_0,x_1)}|u^{(4)}(x)|$$

$$+\frac{C(\Delta x)^2}{\Gamma(4-\alpha)}(x_j - x_0)^{1-\alpha}\max_{x\in(x_0,x_1)}|u^{(3)}(x)|.$$

$\square$

Lemma 5.3 *For $j = 2, \ldots, N - 1$, define*

$$\epsilon_{2,4}(x_j) = \frac{1}{(\Delta x)^2}\frac{1}{\Gamma(4-\alpha)}\sum_{p=0}^{2}(-1)^p\binom{2}{p}$$

$$\times\left(\sum_{k=1}^{j-1+p}\int_{x_{k-1}}^{x_k} l_{k,4}(\xi)u^{(4)}(\xi_k)(x_{j-1+p} - \xi)^{1-\alpha}d\xi\right),$$

for $\xi_k \in (x_{k-1}, x_k)$ and with $|l_{k,4}(\xi)| \leq C(\Delta x)^4$, where C is a positive constant that do not depend on k and Δx. Then we have the local bound

$$|\epsilon_{2,4}(x_j)| \leq \frac{4C(\Delta x)^2}{(2-\alpha)\Gamma(4-\alpha)}\max_{x\in(x_0,x_{j+1})}|u^{(4)}(x)|(x_j - x_0)^{2-\alpha}$$

$$+\frac{C(\Delta x)^{4-\alpha}}{(2-\alpha)\Gamma(4-\alpha)}\max_{x\in(x_0,x_{j+1})}|u^{(4)}(x)|.$$

Proof For $\epsilon_{2,4}(x_j)$ we have

$$|\epsilon_{2,4}(x_j)| \leq \frac{C(\Delta x)^2}{(2-\alpha)\Gamma(4-\alpha)}\max_{x\in(x_0,x_{j+1})}|u^{(4)}(x)|$$

$$\times\left((x_{j+1} - x_0)^{2-\alpha} + 2(x_j - x_0)^{2-\alpha} + (x_{j-1} - x_0)^{2-\alpha}\right),$$

that is,

$$|\epsilon_{2,4}(x_j)| \leq \frac{4C(\Delta x)^2}{(2-\alpha)\Gamma(4-\alpha)}\max_{x\in(x_0,x_{j+1})}|u^{(4)}(x)|(x_j - x_0)^{2-\alpha}$$

$$+\frac{C(\Delta x)^{4-\alpha}}{(2-\alpha)\Gamma(4-\alpha)}\max_{x\in(x_0,x_{j+1})}|u^{(4)}(x)|.$$

$\square$

The main result can now be established, using the preceding lemmas.

Theorem 5.5 *Let* $u \in AC^4([a,b])$ *and* $u^{(4)} \in L_\infty(a,b)$ *and* $\epsilon_T(\Delta x, \alpha, a, x_j)$ *defined by (5.21). Then, for* $x_0 = a$ *and* Δx *small enough, there exists* $\eta_j \in (x_{j-1}, x_{j+1})$, *for* $j = 1, \ldots, N-1$, *such that*

$$|\epsilon_T(\Delta x, \alpha, a, x_j)| \le c_{\alpha,1}(\Delta x)^2 (x_j - x_0)^{-\alpha} \max_{x \in (x_0, x_1)} |u^{(2)}(x)|$$

$$+ c_{\alpha,2}(\Delta x)^2 (x_j - x_0)^{1-\alpha} \max_{x \in (x_0, x_1)} |u^{(3)}(x)|$$

$$+ c_{\alpha,3}(\Delta x)^2 (x_j - x_0)^{2-\alpha} \max_{x \in (x_0, x_{j+1})} |u^{(4)}(x)|$$

$$+ c_{\alpha,4}(\Delta x)^{4-\alpha} \max_{x \in (x_0, x_{j+1})} |u^{(4)}(x)| + c_{\alpha,5}(\Delta x)^2 \left| D_{a+}^{\alpha+2} u(\eta_j) \right|,$$

where $c_{\alpha,k}$, $k = 1, \ldots, 5$ *are positive constants that only depend on* α *and the fractional operator on the right hand side is defined according to (5.16).*

Proof Proceeding in an analogous manner to the proof outlined for Theorem 4.13, we can establish that, for $j = 1, \ldots, N-1$, there exists an $\eta_j \in (x_{j-1}, x_{j+1})$, such that,

$$\frac{d^2}{dx^2} I_{a+}^{2-\alpha} u(x_j) = \frac{I_{a+}^{2-\alpha} u(x_{j+1}) - 2 I_{a+}^{2-\alpha} u(x_j) + I_{a+}^{2-\alpha} u(x_{j-1})}{(\Delta x)^2} + \epsilon_1(\eta_j),$$

where

$$\epsilon_1(\eta_j) = -\frac{(\Delta x)^2}{12} D_{a+}^{\alpha+2} u(\eta_j). \tag{5.25}$$

This fractional derivative of order $\alpha + 2$ is defined according to (5.16).

A second error is related to the approximation of the fractional integral, that is,

$$\frac{d^2}{dx^2} I_{a+}^{2-\alpha} u(x_j) = \frac{1}{(\Delta x)^2} \left(\tilde{I}_{a+}^{2-\alpha} u(x_{j+1}) - 2 \tilde{I}_{a+}^{2-\alpha} u(x_j) + \tilde{I}_{a+}^{2-\alpha} u(x_{j-1}) \right)$$

$$+ \epsilon_2(x_j) + \epsilon_1(\eta_j), \tag{5.26}$$

where

$$\epsilon_2(x_j) = \frac{1}{(\Delta x)^2} \sum_{p=0}^{2} (-1)^p \binom{2}{p} \left[I_{a+}^{2-\alpha} u(x_{j-1+p}) - \tilde{I}_{a+}^{2-\alpha} u(x_{j-1+p}) \right]. \tag{5.27}$$

We need to discuss the approximation of the fractional integral at each point x_{j+1}, x_j, x_{j-1} involved in the error denoted by ϵ_2. The initial steps of the proof to found an upper bound for $\epsilon_2(x_j)$ are similar to the proof of Theorem 4.13 in

Sect. 4.3.1. Then, the proof differs substantially in the approach we need to do at the nearest interval to the boundary, that is, in (x_0, x_1).

We have

$$I_{a+}^{2-\alpha} u(x_j) - \tilde{I}_{a+}^{2-\alpha} u(x_j) = \frac{1}{\Gamma(4-\alpha)} \sum_{k=1}^{j} \int_{x_{k-1}}^{x_k} (u(\xi) - s_k(\xi))(x_j - \xi)^{1-\alpha} d\xi.$$

(5.28)

Therefore, for $j \geq 2$,

$$\epsilon_2(x_j) = \frac{1}{(\Delta x)^2} \frac{1}{\Gamma(4-\alpha)} \sum_{p=0}^{2} (-1)^p \binom{2}{p}$$
$$\times \left(\sum_{k=1}^{j-1+p} \int_{x_{k-1}}^{x_k} (u(\xi) - s_k(\xi))(x_{j-1+p} - \xi)^{1-\alpha} d\xi \right)$$

and for $j = 1$ we have only two terms on the first sum, that is,

$$\epsilon_2(x_j) = \frac{1}{(\Delta x)^2} \frac{1}{\Gamma(4-\alpha)} \sum_{p=1}^{2} (-1)^p \binom{2}{p}$$
$$\times \left(\sum_{k=1}^{j-1+p} \int_{x_{k-1}}^{x_k} (u(\xi) - s_k(\xi))(x_{j-1+p} - \xi)^{1-\alpha} d\xi \right).$$

Taking in consideration Lemma 4.1, we have

$$\epsilon_2(x_j) = - \sum_{r=2}^{4} \frac{1}{r!} \epsilon_{2,r}(x_j),$$

where, for $r = 2, 3$

$$\epsilon_{2,r}(x_j) = \frac{1}{(\Delta x)^2} \frac{1}{\Gamma(4-\alpha)} \sum_{p=0}^{2} (-1)^p \binom{2}{p}$$
$$\times \left(\sum_{k=1}^{j-1+p} \int_{x_{k-1}}^{x_k} l_{k,r}(\xi) u^{(r)}(\xi)(x_{j-1+p} - \xi)^{1-\alpha} d\xi \right)$$

and for $r = 4$ we have

$$\epsilon_{2,4}(x_j) = \frac{1}{(\Delta x)^2} \frac{1}{\Gamma(4-\alpha)} \sum_{p=0}^{2} (-1)^p \binom{2}{p}$$

$$\times \left(\sum_{k=1}^{j-1+p} \int_{x_{k-1}}^{x_k} l_{k,4}(\xi) u^{(4)}(\xi_k)(x_{j-1+p} - \xi)^{1-\alpha} d\xi \right).$$

We can proceed in a similar manner for $j = 1$ and in this case we do not have the term $p = 0$, associated to the point x_{j-1}, on the expressions of $\epsilon_{2,r}(x_j)$, $r = 2, 3, 4$.

By Lemma 5.1, Lemma 5.2 and Lemma 5.3, we have, respectively, for $j \geq 2$,

$$|\epsilon_{2,2}(x_j)| \leq \frac{C_2(\Delta x)^2}{(2-\alpha)\Gamma(4-\alpha)}(x_j - x_0)^{2-\alpha} \max_{x\in(x_2,x_j)} |u^{(4)}(x)|$$

$$+ \frac{C_2 2^{\alpha-1}(\Delta x)^2}{\Gamma(4-\alpha)}(x_j - x_0)^{1-\alpha} \max_{x\in(x_0,x_1)} |u^{(3)}(x)|$$

$$+ \frac{C_2(\Delta x)^2}{\Gamma(4-\alpha)}(x_j - x_0)^{-\alpha} \max_{x\in(x_0,x_1)} |u^{(2)}(x)|,$$

$$|\epsilon_{2,3}(x_j)| \leq \frac{C_3(\Delta x)^2}{(2-\alpha)\Gamma(4-\alpha)}(x_j - x_0)^{2-\alpha} \max_{x\in(x_2,x_j)} |u^{(4)}(x)|$$

$$+ \frac{2C_3(\Delta x)^2}{(2-\alpha)\Gamma(4-\alpha)}(x_j - x_0)^{2-\alpha} \max_{x\in(x_0,x_1)} |u^{(4)}(x)|$$

$$+ \frac{C_3(\Delta x)^2}{\Gamma(4-\alpha)}(x_j - x_0)^{1-\alpha} \max_{x\in(x_0,x_1)} |u^{(3)}(x)|,$$

$$|\epsilon_{2,4}(x_j)| \leq \frac{4C_4(\Delta x)^2}{(2-\alpha)\Gamma(4-\alpha)} \max_{x\in(x_0,x_{j+1})} |u^{(4)}(x)|(x_j - x_0)^{2-\alpha}$$

$$+ \frac{C_4(\Delta x)^{4-\alpha}}{(2-\alpha)\Gamma(4-\alpha)} \max_{x\in(x_0,x_{j+1})} |u^{(4)}(x)|.$$

For $j = 1$, taking into account the proof of Lemma 5.1, $\epsilon_{2,2}(x_1) = A_1$, that is,

$$\epsilon_{2,2}(x_1) = \int_{x_0}^{x_1} l_{1,2}(\xi) u^{(2)}(\xi) \left((x_2 - \xi)^{1-\alpha} - (x_1 - \xi)^{1-\alpha} \right) d\xi$$

$$+ \Delta x \int_{x_0}^{x_1} l_{1,2}(\xi) u^{(3)}(\xi_1)(x_1 - \xi)^{1-\alpha} d\xi$$

and similarly we can obtain

$$|\epsilon_{2,2}(x_1)| \leq \frac{C_2 2^{\alpha-1}(\Delta x)^2}{\Gamma(4-\alpha)}(x_1 - x_0)^{1-\alpha} \max_{x\in(x_0,x_1)} |u^{(3)}(x)|$$

$$+ \frac{C_2(\Delta x)^2}{\Gamma(4-\alpha)}(x_1 - x_0)^{-\alpha} \max_{x\in(x_0,x_1)} |u^{(2)}(x)|.$$

To obtain a bound for $\epsilon_{2,3}(x_1)$ and $\epsilon_{2,4}(x_1)$, we proceed in a similar way taking into account the proof of Lemma 5.2 and Lemma 5.3 respectively.

Finally, we have,

$$|\epsilon_2(x_j)| \leq c_{\alpha,1}(\Delta x)^2(x_j - x_0)^{-\alpha} \max_{x\in(x_0,x_1)} |u^{(2)}(x)|$$

$$+ c_{\alpha,2}(\Delta x)^2(x_j - x_0)^{1-\alpha} \max_{x\in(x_0,x_1)} |u^{(3)}(x)|$$

$$+ c_{\alpha,3}(\Delta x)^2(x_j - x_0)^{2-\alpha} \max_{x\in(x_0,x_{j+1})} |u^{(4)}(x)|$$

$$+ c_{\alpha,4}(\Delta x)^{4-\alpha} \max_{x\in(x_0,x_{j+1})} |u^{(4)}(x)|.$$

The result stated in the theorem follows from here. $\qquad\square$

From the previous result we can derive the following statement, that reinforces that consistency depends on the value of the initial point of the domain.

Corollary 5.1 *Let $u \in AC^4([a, b])$ and $u^{(4)} \in L_\infty(a, b)$ and $\epsilon_T(\Delta x, \alpha, a, x_j)$ defined by (5.21). Then, for $x_0 = a$ and Δx small enough, we have, for $j = 1, \ldots, N - 1$,*

$$|\epsilon_T(\Delta x, \alpha, a, x_j)| \leq (\Delta x)^2 \sum_{k=0}^{3} d_{\alpha,k}(x_j - x_0)^{k-\alpha-2}|u^{(k)}(x_0)| + (\Delta x)^2 d_{\alpha,4}||u^{(4)}||_\infty,$$

where $d_{\alpha,k}$, $k = 0, \ldots, 4$ are positive constants that only depend on α.

Proof This result follows directly from the previous theorem by observing two main properties.

First, for $x \in (x_0, x_1)$, there exists $0 < p < 1$, such that, we have $u^{(k)}(x) = u^{(k)}(x_0 + p\Delta x)$, $k = 2, 3$ and therefore $u^{(k)}(x_0 + p\Delta x) = u^{(k)}(x_0) + p\Delta x u^{(k)}(\xi)$.

Secondly, since $u \in AC^4([a, b])$, then [119, p. 39],

$$D_{a+}^{(\alpha+2)}u(x_j) = \sum_{k=0}^{3} \frac{u^{(k)}(x_0)}{\Gamma(k-\alpha-1)}(x_j - x_0)^{k-\alpha-2}$$

$$+ \frac{1}{\Gamma(2-\alpha)} \int_{x_0}^{x_j} u^{(4)}(\xi)(x_j - \xi)^{-\alpha+1}d\xi.$$

Therefore, since $u^{(4)} \in L_\infty([a, b])$, it follows

$$\left| D_{a+}^{(\alpha+2)} u(x_j) \right| \leq \sum_{k=0}^{3} c_{\alpha,k} (x_j - x_0)^{k-\alpha-2} |u^{(k)}(x_0)| + c_{\alpha,4} ||u^{(4)}||_\infty,$$

where $c_{\alpha,k}$, $k = 0, \dots, 4$ are positive constants that only depend on α. $\square$

From the previous result we infer that to attain second order accuracy it is necessary to have the derivatives, up to the third order, zero at the initial point of the domain. Otherwise the order of accuracy will be dominated by the local truncation error at the nearest point to the boundary.

What happens if the solutions have derivatives unbounded, that is, if they behave near the boundary as $(x - a)^\gamma$?

Without loss of generality consider $a = 0$. Assume the function near the boundary is of the form

$$u(x) = x^\gamma, \qquad \gamma - \alpha < 2.$$

The nearest point to the boundary is $x_1 = \Delta x$.

Take the error (5.26) in Theorem 5.5 to discuss what happens at x_1. The leading error term is given by

$$\epsilon^{lead}(x_1) = \epsilon_1^{lead}(x_1) + \epsilon_2^{lead}(x_1),$$

where $\epsilon_1^{lead}(x_1)$ is defined according to (5.25) and $\epsilon_2^{lead}(x_1)$ is defined according to (5.27).

The error $\epsilon_1^{lead}(x_1)$ is dominated by

$$(\Delta x)^2 D_{0+}^{\alpha+2} u(\eta_1), \quad \text{for} \quad 0 < \eta_1 < \Delta x,$$

that is, is dominated by

$$(\Delta x)^2 (\Delta x)^{\gamma-\alpha-2} = (\Delta x)^{\gamma-\alpha}.$$

Now consider the error $\epsilon_2^{lead}(x_1)$ dominated by the first integral in the formulation (5.28). The dominating factor is

$$I_1 = \frac{1}{(\Delta x)^2} \int_0^{\Delta x} (u(\xi) - s_1(\xi))(x_1 - \xi)^{1-\alpha} d\xi,$$

where

$$s_1(\xi) = \frac{x_1 - \xi}{\Delta x} u(x_0) + \frac{\xi - x_0}{\Delta x} u(x_1) = \xi(\Delta x)^{\gamma-1}.$$

Therefore

$$I_1 = \frac{1}{(\Delta x)^2} \int_0^{\Delta x} (\xi^\gamma - \xi(\Delta x)^{\gamma-1})(\Delta x - \xi)^{1-\alpha} d\xi.$$

Hence

$$|I_1| \leq \frac{(\Delta x)^\gamma}{(\Delta x)^2} \int_0^{\Delta x} (\Delta x - \xi)^{1-\alpha} d\xi$$

and $\epsilon_2^{lead}(x_1)$ behaves also according to $(\Delta x)^{\gamma-\alpha}$.

We can formulate the following result.

Corollary 5.2 *Let $u \in AC([a,b]) \cap C^4(a,b]$ and u near the left boundary behaves as $(x-a)^\gamma$, $\gamma > 0$, that is,*

$$\sup_{x \in (a,b)} (x-a)^{n-\gamma} |u^{(n)}(x)| < \infty, \ n = 2, 3, 4.$$

The error $\epsilon_T(\Delta x, \alpha, a, x_j)$ defined by (5.21) is such that,

$$|\epsilon_T(\Delta x, \alpha, a, x_j)| \leq C_\alpha (\Delta x)^p, \ p = \min\{2, \gamma - \alpha\},$$

where C_α is a positive constant that only depends on α.

5.3   Numerical Methods and its Consistency

The numerical methods can be similar to the ones we have discussed in the previous chapter. The main change is related to the fractional operator that is defined in a bounded domain, instead of being defined in the real line. Specifically the fractional integral has changed from starting at $-\infty$ to start at the point where the boundary is located. For simplicity let us assume that the location of the absorbing boundary is at $x = 0$.

Suppose we choose to approximate the fractional derivative by the approximation which we have proved it is second order accurate when the problem is defined in the real line. We have seen in Sect. 5.2 that the accuracy can be of lower order when the problem is defined in a bounded domain, depending not necessarily on the regularity of the function but on the values of the function and its derivatives at the initial point. A similar behaviour has been reported in [137], for a lower order approximation of the fractional derivative, and it has also been described in Sect. 5.2. This lost of accuracy will be reflected in the order of consistency of the full discretisation numerical method.

To find an approximate solution for the problem (5.4)–(5.7), consider the uniform discretisation in time $t_{n+1} = t_n + \Delta t$, where Δt is the time step. We denote by U_j^n the approximation to the solution $u(x_j, t_n)$.

Based on what has been discussed in Sect. 5.2, define the discrete operator that approximates the left fractional Riemann-Liouville derivative by

$$\delta_{l,2,abs}^\alpha U_j^n = \frac{1}{\Gamma(4-\alpha)} \sum_{m=-1}^{j} q_m^\alpha U_{j-m}^n. \tag{5.29}$$

The Crank-Nicolson method, to solve Eq. (5.4), is given by

$$\frac{U_j^{n+1} - U_j^n}{\Delta t} = \frac{D}{2(\Delta x)^\alpha} \delta_{l,2,abs}^\alpha U_j^n + \frac{D}{2(\Delta x)^\alpha} \delta_{l,2,abs}^\alpha U_j^{n+1} + f_j^{n+1/2}, \tag{5.30}$$

where $f_j^{n+1/2} = (f_j^{n+1} + f_j^n)/2$.

Let

$$\mu_\alpha = \frac{D\Delta t}{(\Delta x)^\alpha}.$$

Then, the numerical method can be re-written as

$$\left(1 - \frac{1}{2}\mu_\alpha \delta_{l,2,abs}^\alpha \right) U_j^{n+1} = \left(1 + \frac{1}{2}\mu_\alpha \delta_{l,2,abs}^\alpha \right) U_j^n + f_j^{n+1/2}. \tag{5.31}$$

It is known that for sufficiently smooth initial data, the Crank-Nicolson discretization attains a second-order accuracy in time. However, the presence of an absorbing boundary condition affects the accuracy of the spatial discretisation and consequently the accuracy of the overall method.

In order to state the truncation error of the numerical method we denote the truncation error of the fractional partial derivative as

$$\epsilon_T(\Delta x, \alpha, a, x_j, t_n) = \frac{\partial^\alpha u}{\partial x^\alpha}(x_j, t_n) - \frac{\delta_{l,2,abs}^\alpha u}{\Delta x^\alpha}(x_j, t_n). \tag{5.32}$$

Corollary 5.3 (Consistency) *Let* $u(x, \cdot) \in C^3(0, \infty)$, $u(\cdot, t) \in AC([a, b]) \cap C^4(a, b]$ *and u near the left boundary behaves as* $(x-a)^\gamma$, $\gamma > 0$, *that is,*

$$\sup_{x \in (a,b)} (x-a)^{n-\gamma} \left| \frac{\partial^n u}{\partial x^n}(x, t) \right| < \infty, \quad n = 2, 3, 4.$$

Then, the local truncation error of the numerical method is given by

$$T_j^n = \frac{(\Delta t)^2}{2} \frac{\partial^3 u(x_j, \tau_n)}{\partial t^3} - \frac{1}{2}\left(\epsilon_T(\Delta x, \alpha, a, x_j, t_{n+1}) + \epsilon_T(\Delta x, \alpha, a, x_j, t_n)\right)$$

for a τ_n *such that* $t_n \leq \tau_n \leq t_{n+1}$ *and* ϵ_T *denotes the truncation error* (5.32).

Proof See Exercise 5.6.   □

The next result follows easily from the result above.

Corollary 5.4 (Order of Accuracy) *Let* $u(x, \cdot) \in C^3(0, \infty)$, $u(\cdot, t) \in AC([a, b]) \cap C^4(a, b]$ *and u near the left boundary behaves as* $(x - a)^\gamma$, $\gamma > 0$, *that is,*

$$\sup_{x \in (a,b)} (x - a)^{n-\gamma} \left| \frac{\partial^n u}{\partial x^n}(x, t) \right| < \infty, \quad n = 2, 3, 4.$$

Then, the global truncation error of the numerical method satisfies

$$\|T^n\|_\infty = \mathscr{O}((\Delta t)^2) + \mathscr{O}((\Delta x)^p), \qquad \|T^n\|_2 = \mathscr{O}((\Delta t)^2) + \mathscr{O}((\Delta x)^q),$$

with $p = \min\{2, \gamma - \alpha\}$ *and* $q = \min\{2, \gamma - \alpha + 0.5\}$.

Other numerical methods can be considered based on what we have discussed previously and using the lower order approximation for the fractional order derivative. Therefore, for a complementar analysis, in the next section, to discuss the stability analysis, we present numerical methods based in the lower order approximation of the fractional derivative.

5.4   Matrix Analysis for Stability

In this section, we discuss the stability of the numerical methods. The von Neumann analysis is not the natural approach to study the stability of the numerical methods, when we have a boundary condition. However, it is known that the von Neumann stability conditions are necessary conditions for the stability of a numerical method for a problem with boundaries. This means that the stability region of the numerical method with boundary conditions is not larger than the stability region of the numerical method without boundary conditions.

For non-local problems, we have an additional detail not present in the classical problems. The formulation of the differential equation is affected by the inclusion of the boundary. Hence, the numerical method will be instantly changed by the fact that the fractional operator is not the same and will behave differently near the boundary points.

We have seen, in Chap. 3, that we can use two different tools to analyse the stability. One is the energy method that will be discussed in the next section and the other one is the matrix analysis that we will discuss in this section. Hence, by analysing the eigenvalues and the norm of the iterative matrix of the numerical methods we are going to determine the stability conditions. To this end, we consider the simplified problem that only involves the left Riemann-Liouville fractional derivative.

Consider Eq. (5.4) without source term

$$\frac{\partial u}{\partial t}(x, t) = \frac{\partial_{abs}^{\alpha} u}{\partial x^{\alpha}}(x, t), \quad x \in (0, b)$$

and with absorbing boundary conditions

$$u(0, t) = u(b, t) = 0.$$

One approximation for the left Riemann-Liouville fractional derivative is based in the Grünwald-Letnikov approximation given by

$$\frac{1}{(\Delta x)^{\alpha}} \sum_{k=0}^{j+1} g_k^{\alpha} U_{j-k+1}^n, \qquad g_k^{\alpha} = (-1)^k \binom{\alpha}{k}.$$

Let us define

$$\delta_{l,1,abs}^{\alpha} U_j^n := \sum_{k=0}^{j+1} g_k^{\alpha} U_{j-k+1}^n.$$

The explicit Euler numerical method for the problem under discussion reads

$$U_j^{n+1} = U_j^n + \mu_\alpha \delta_{l,1,abs}^{\alpha} U_j^n. \tag{5.33}$$

According to the stability result of the previous chapter, if the problem is defined in the real line a necessary stability condition for this numerical method is

$$\mu_\alpha \leq 2^{1-\alpha}.$$

In the matricial form we have the numerical method

$$U^{n+1} = (I + \mu_\alpha L)U^n + b^n,$$

where

$$L = \begin{bmatrix} g_1^{\alpha} & g_0^{\alpha} & 0 & \cdots & & 0 \\ g_2^{\alpha} & & & & & \\ & & \ddots & & \vdots & \\ \vdots & & & \ddots & & 0 \\ & & & & & g_0^{\alpha} \\ g_{N-1}^{\alpha} & \cdots & & & g_2^{\alpha} & g_1^{\alpha} \end{bmatrix}, \quad U^n = \begin{bmatrix} U_1^n \\ U_2^n \\ \vdots \\ U_{N-1}^n \end{bmatrix}, \quad b^n = \begin{bmatrix} 0 \\ \vdots \\ 0 \\ g_0^{\alpha} U_N^n \end{bmatrix}.$$

In the next result we compute the spectral radius of the iterative matrix of the numerical method, by using the Gershgorin theorem. A similar discussion has been done in [74].

Proposition 5.1 *Let M be the iterative matrix of the Euler explicit scheme in the presence of absorbing boundary conditions. If $\mu_\alpha \leq 1/\alpha$ then $\rho(M) \leq 1$.*

Proof By the Gershgorin theorem we have, for the rows $j = 1, \ldots, N - 2$, that

$$|\lambda - (1 + \mu_\alpha g_1^\alpha)| \leq \mu_\alpha \left(g_0^\alpha + \sum_{k=2}^{j} g_k^\alpha \right).$$

For the last row, $j = N - 1$, we have

$$|\lambda - (1 + \mu_\alpha g_1^\alpha)| \leq \mu_\alpha \left(\sum_{k=2}^{N-1} g_k^\alpha \right).$$

We know that

$$g_0^\alpha + \sum_{k=2}^{j} g_k^\alpha \leq -g_1^\alpha = \alpha, \ \ j = 1, \ldots, N - 2.$$

Therefore

$$|\lambda - (1 + \mu_\alpha g_1^\alpha)| \leq \mu_\alpha \alpha, \ \ j = 1, \ldots, N - 2.$$

Similarly, for $j = N - 1$, since

$$\sum_{k=2}^{j} g_k^\alpha \leq -g_1^\alpha - g_0^\alpha = \alpha - 1 \leq \alpha,$$

then

$$|\lambda - (1 + \mu_\alpha g_1^\alpha)| \leq \mu_\alpha \alpha.$$

Given two complex numbers we have $||z| - |z'|| \leq |z - z'|$. Hence, we obtain

$$||\lambda| - |1 + \mu_\alpha g_1^\alpha|| \leq |\lambda - (1 + \mu_\alpha g_1^\alpha)| \leq \mu_\alpha \alpha.$$

By hypotheses $\mu_\alpha \alpha \leq 1$ and therefore

$$1 + \mu_\alpha g_1^\alpha = 1 - \alpha \mu_\alpha \geq 0.$$

Hence

$$|\lambda| - (1 - \mu_\alpha \alpha) \leq \mu_\alpha \alpha,$$

that is,

$$|\lambda| \leq 1.$$

$\square$

Consider the implicit Euler numerical method, for the problem under discussion,

$$U_j^{n+1} = U_j^n + \mu_\alpha \delta_{l,1,abs}^\alpha U_j^{n+1}. \tag{5.34}$$

The matricial form is

$$(1 - \mu_\alpha L)U^{n+1} = U^n. \tag{5.35}$$

In the next result, it is shown that the numerical method is unconditionally stable, by showing that $||U^{n+1}|| \leq ||U^n||$, where $|| \cdot ||$ stands for the 2-norm.

Proposition 5.2 *Consider the implicit Euler scheme (5.35) in the presence of absorbing boundary conditions. We have $||U^{n+1}|| \leq ||U^n||$.*

Proof The matrix $- L$ is strictly definite positive (see Exercise 5.10). This means that $(1 - \mu_\alpha L)$ has an inverse matrix (see Exercise 5.10) and therefore we can write

$$U^{n+1} = (1 - \mu_\alpha L)^{-1} U^n.$$

We also know (see Exercise 5.11) that for the 2-nom

$$||(1 - \mu_\alpha L)^{-1}|| \leq 1.$$

Therefore

$$||U^{n+1}|| = ||(1 - \mu_\alpha L)^{-1} U^n|| \leq ||(1 - \mu_\alpha L)^{-1}|| ||U^n|| \leq ||U^n||.$$

$\square$

Consider the Crank-Nicolson numerical method, for the same equation as previously, given by

$$\left(1 - \frac{1}{2}\mu_\alpha \delta_{l,1,abs}^\alpha\right) U_j^{n+1} = \left(1 + \frac{1}{2}\mu_\alpha \delta_{l,1,abs}^\alpha\right) U_j^n. \tag{5.36}$$

The matricial form is now

$$\left(1 - \frac{1}{2}\mu_\alpha L\right) U^{n+1} = \left(1 + \frac{1}{2}\mu_\alpha L\right) U^n. \tag{5.37}$$

Let us prove that we also have an unconditionally stable method.

Proposition 5.3 *Consider the Crank Nicolson method scheme (5.37) in the presence of absorbing boundary conditions. We have $\|U^{n+1}\| \le \|U^n\|$.*

Proof This proof also follows from the result that is given as an exercise (see Exercise 5.10). It states that if $-L$ is definite positive, for $\theta \ge 0$ then

$$\|(I - \theta L)^{-1}(I + \theta L)\| \le 1.$$

From the matricial form we have

$$U^{n+1} = \left(1 - \frac{1}{2}\mu_\alpha L\right)^{-1}\left(1 + \frac{1}{2}\mu_\alpha L\right)U^n.$$

Therefore

$$\|U^{n+1}\| = \left\|\left(1 - \frac{1}{2}\mu_\alpha L\right)^{-1}\left(1 + \frac{1}{2}\mu_\alpha L\right)U^n\right\|$$

$$\le \left\|\left(1 - \frac{1}{2}\mu_\alpha L\right)^{-1}\left(1 + \frac{1}{2}\mu_\alpha L\right)\right\|\,\|U^n\|$$

$$\le \|U^n\|.$$

$\square$

A similar analysis can be done for numerical methods based on a different approximation of the fractional derivative, namely based on the approximation derived using the integral form, which was second order in the real line. However, it may not be a straightforward calculation from what we have presented here, because the sign of the coefficients of the quadrature formula that approximates the fractional derivative can be different. This detail may change the strategy of the proof. In the next section we show how to study the stability analysis based in the energy method and we pay attention to the second order approximation of the fractional derivative, which has not been considered in this section.

The stability of the numerical method only guaranties convergence for consistent schemes. We have seen that for some type of solutions the numerical methods presented may be inconsistent. Therefore, in the next section, we show how to use the energy method to prove that an inconsistent scheme can converge.

5.5   The Energy Method: Stability and Convergence

In this section, we consider a numerical method that consists on the Crank-Nicolson method for the time discretisation and for the spatial discretisation the approximation of the fractional derivative is based on the linear spline approximation of the solution.

Consider the equation

$$\frac{\partial u}{\partial t}(x,t) = \frac{\partial_{abs}^{\alpha} u}{\partial x^{\alpha}}(x,t), \quad x \in (0, b),$$

with an initial condition and absorbing boundary conditions

$$u(0, t) = u(b, t) = 0.$$

The left Riemann-Liouville fractional derivative is approximated by

$$\frac{1}{(\Delta x)^{\alpha} \Gamma(4 - \alpha)} \sum_{k=-1}^{j} q_k^{\alpha} U_{j-k}^n.$$

Define

$$\delta_{l,2,abs}^{\alpha} U_j^n = \frac{1}{\Gamma(4 - \alpha)} \sum_{k=-1}^{j} q_k^{\alpha} U_{j-k}^n.$$

The Crank-Nicolson numerical method reads

$$\left(1 - \frac{1}{2}\mu_\alpha \delta_{l,2,abs}^{\alpha}\right) U_j^{n+1} = \left(1 + \frac{1}{2}\mu_\alpha \delta_{l,2,abs}^{\alpha}\right) U_j^n. \tag{5.38}$$

In the next subsections we discuss the stability and convergence for this numerical method.

5.5.1 Stability

We start with some properties required to prove the main result on the stability. The next lemma presents properties of the coefficients (5.20) that appear on the approximation of the Riemann-Liouville fractional derivative. Some other properties have been already given in Lemma 4.2 of the previous chapter.

Lemma 5.4 *The coefficients q_m^{α} defined by (5.20) satisfy the following condition: For a fixed positive p, let $s_p := \sum_{k=-1}^{p} q_k^{\alpha}$. An upper bound for this finite sum is given by $s_p < -C_\alpha p^{-\alpha}$, for a positive constant C_α.*

Proof We first note that

$$s_p = -a_p^{\alpha} + a_{p+1}^{\alpha},$$

where the a_p^α's are defined in (5.19). This means

$$s_p = (p+2)^{3-\alpha} - 3(p+1)^{3-\alpha} + 3p^{3-\alpha} - (p-1)^{3-\alpha}.$$

By writing each term in its binomial expansion, we can obtain

$$s_p = p^{3-\alpha} \sum_{k=3}^{\infty} \binom{3-\alpha}{k} \frac{1}{p^k}(2^k - 3 - (-1)^k).$$

We know that, see [119, p. 14],

$$\binom{3-\alpha}{k} = \frac{(-1)^{k-1}\alpha\Gamma(k-\alpha)}{\Gamma(1-\alpha)\Gamma(k+1)}$$

and therefore

$$s_p = \frac{1}{p^\alpha} \frac{\alpha}{\Gamma(1-\alpha)} B_\alpha,$$

where

$$B_\alpha = \sum_{k=3}^{\infty} (-1)^{k-1} \left(\frac{1}{p}\right)^{k-3} \frac{\Gamma(k-\alpha)}{\Gamma(k+1)}(2^k - 3 - (-1)^k).$$

Since B_α is an alternating series, it is easy to conclude that $B_\alpha > \Gamma(3-\alpha)$. By noting that $\Gamma(1-\alpha) < 0$, for $\alpha \in (1, 2)$, we can conclude that

$$s_p < -\frac{1}{p^\alpha} \frac{\alpha(-\Gamma(3-\alpha))}{\Gamma(1-\alpha)}.$$

The stated result follows for $C_\alpha = -\alpha\Gamma(3-\alpha)/\Gamma(1-\alpha) > 0$. $\square$

Let $u_0 = 0$ and $u_N = 0$ and define the discrete inner product and its norm

$$(u, v) = \Delta x \sum_{j=1}^{N-1} u_i v_i, \quad ||u|| = \sqrt{(u, u)}, \quad u, v \in \mathbb{R}^{N-1}.$$

The next result is needed to prove the stability in the L_2 discrete norm.

Lemma 5.5 *Let* $\mathbf{e} = (e_1, \ldots, e_{N-1})$ *with* $e_0 = e_N = 0$ *and*

$$\delta_{l,2,abs}^\alpha e_j = \frac{1}{\Gamma(4-\alpha)} \sum_{k=1}^{j+1} q_{j-k}^\alpha e_k, \quad j = 1, \ldots, N-1.$$

Then

$$\frac{1}{(\Delta x)^\alpha}(\delta^\alpha_{l,2,abs}\mathbf{e}, \mathbf{e}) \le -C_\alpha(x_N)^{-\alpha}||\mathbf{e}||^2,$$

where C_α is a positive constant that only depends on α.

Proof We have that

$$\left(\delta^\alpha_{l,2,abs}\mathbf{e}, \mathbf{e}\right) = \sum_{j=1}^{N-1} \Delta x \delta^l e_j e_j.$$

We can obtain successively

$$\sum_{j=1}^{N-1} \delta^\alpha_{l,2,abs} e_j e_j = \frac{1}{\Gamma(4-\alpha)} \sum_{j=1}^{N-1} \left(\sum_{k=1}^{j+1} q^\alpha_{j-k} e_k\right) e_j$$

$$= \frac{1}{\Gamma(4-\alpha)} \left\{ q^\alpha_0 ||\mathbf{e}||^2 + (q^\alpha_1 + q^\alpha_{-1}) \sum_{j=1}^{N-2} e_j e_{j+1} + q^\alpha_2 \sum_{j=1}^{N-3} e_j e_{j+2} \right.$$

$$\left. + \cdots + q^\alpha_{N-2} \sum_{j=1}^{1} e_j e_{j+N-3} \right\}.$$

This can be written as

$$\sum_{j=1}^{N-1} \delta^\alpha_{l,2,abs} e_j e_j$$

$$= \frac{1}{\Gamma(4-\alpha)} q^\alpha_0 ||\mathbf{e}||^2 + \frac{1}{\Gamma(4-\alpha)}$$

$$\times \left\{ (q^\alpha_1 + q^\alpha_{-1}) \sum_{j=1}^{N-2} e_j e_{j+1} + \sum_{k=2}^{N-2} q^\alpha_k \left(\sum_{j=1}^{N-(k+1)} e_j e_{j+k}\right) \right\}.$$

From the properties of the coefficients q^α_m that have been stated in Lemma 5.2 of the previous chapter, we know that $q^\alpha_m \ge 0$, for $m \ge 2$, and $q^\alpha_1 + q^\alpha_{-1} \ge 0$. Using the fact that

$$\sum_{j=1}^{N-(k+1)} \Delta x e_j e_{j+k} \le ||\mathbf{e}||^2, \quad k = 1, 2, \ldots, N-2,$$

we have

$$\frac{1}{(\Delta x)^\alpha} \left(\delta^\alpha_{l,2,abs} \mathbf{e}, \mathbf{e} \right) \le \frac{1}{\Gamma(4-\alpha)} \frac{1}{(\Delta x)^\alpha} \left(\sum_{k=-1}^{N-2} q_k^\alpha \right) ||\mathbf{e}||^2.$$

By the previous lemma we know that $s_{N-2} \le -C_\alpha (N-2)^{-\alpha}$.
Therefore, since $1/N < 1/(N-2)$ it follows that

$$\frac{1}{(\Delta x)^\alpha} \left(\delta^\alpha_{l,2,abs} \mathbf{e}, \mathbf{e} \right) \le -C_\alpha (x_N)^{-\alpha} ||\mathbf{e}||^2.$$

$\square$

The previous lemma is necessary to prove the next stability theorem.

Theorem 5.6 (Stability) *Let $U^k = (U_1^k, \ldots, U_{N-1}^k)$ be a solution of the numerical method (5.31), for a given initial condition $U^0 = (U_1^0, \ldots, U_{N-1}^0)$ and the boundary conditions U_0^k and U_N^k are equal to zero. Then*

$$\max_{0 \le k \le M} ||U^k|| \le \left[||U^0||^2 + 2\epsilon \sum_{n=0}^{M-1} \Delta t ||f^{n+1/2}||^2 \right]^{1/2}.$$

Proof From (5.31) we have

$$\left(\frac{U^{n+1} - U^n}{\Delta t}, \frac{U^{n+1} + U^n}{2} \right) = \frac{D}{(\Delta x)^\alpha} \left(\delta^\alpha_{l,2,abs} \left(\frac{U^{n+1} + U^n}{2} \right), \frac{U^{n+1} + U^n}{2} \right)$$
$$+ \left(f^{n+1/2}, \frac{U^{n+1} + U^n}{2} \right),$$

that is,

$$\frac{1}{2\Delta t}(||U^{n+1}||^2 - ||U^n||^2) = \frac{D}{(\Delta x)^\alpha} \left(\delta^\alpha_{l,2,abs} \left(\frac{U^{n+1} + U^n}{2} \right), \frac{U^{n+1} + U^n}{2} \right)$$
$$+ \left(f^{n+1/2}, \frac{U^{n+1} + U^n}{2} \right).$$

Let $U^{n+1/2} = (U^{n+1} + U^n)/2$. By the Young's inequality, for all $\epsilon > 0$, we have

$$\left(f^{n+1/2}, U^{n+1/2} \right) \le \epsilon ||f^{n+1/2}||^2 + \frac{1}{4\epsilon} ||U^{n+1/2}||^2$$

and therefore

$$\frac{1}{2\Delta t}\left(||U^{n+1}||^2 - ||U^n||^2\right) \leq \frac{D}{(\Delta x)^\alpha}\left(\delta^\alpha_{l,2,abs}\left(U^{n+1/2}\right), U^{n+1/2}\right)$$

$$+\epsilon||f^{n+1/2}||^2 + \frac{1}{4\epsilon}||U^{n+1/2}||^2.$$

In view of the previous lemma, it follows

$$\frac{D}{(\Delta x)^\alpha}\left(\delta^\alpha_{l,2,abs}\left(U^{n+1/2}\right), U^{n+1/2}\right) \leq -C_\alpha(x_N)^{-\alpha}||U^{n+1/2}||^2,$$

for a positive constant C_α. Therefore

$$\frac{1}{2\Delta t}\left(||U^{n+1}||^2 - ||U^n||^2\right) \leq -C_\alpha(x_N)^{-\alpha}||U^{n+1/2}||^2 + \epsilon||f^{n+1/2}||^2 + \frac{1}{4\epsilon}||U^{n+1/2}||^2.$$

Let ϵ be such that

$$-C_\alpha(x_N)^{-\alpha} + \frac{1}{4\epsilon} = 0.$$

Then

$$||U^{n+1}||^2 \leq ||U^n||^2 + 2\epsilon\Delta t||f^{n+1/2}||^2.$$

Upon summation through all $n = 0, 1, \ldots, k$ this implies that

$$||U^k||^2 \leq ||U^0||^2 + 2\epsilon\sum_{n=0}^{k-1}\Delta t||f^{n+1/2}||^2,$$

for all k, $1 \leq k \leq M$. It follows that

$$\max_{0\leq k\leq M}||U^k||^2 \leq ||U^0||^2 + 2\epsilon\sum_{n=0}^{M-1}\Delta t||f^{n+1/2}||^2.$$

Hence

$$\max_{0\leq k\leq M}||U^k|| \leq \left[||U^0||^2 + 2\epsilon\sum_{n=0}^{M-1}\Delta t||f^{n+1/2}||^2\right]^{1/2},$$

which expresses the continuous dependence of the solution of the finite difference scheme on the initial data U^0 and the right-hand side source term f^n. $\square$

In the next section, we present the results on the convergence. We will see that the rate of convergence can be higher than the order of accuracy determined by the consistency analysis.

5.5.2   Convergence

The rate of convergence of the full method can be higher than the order of accuracy of the truncation error and in many cases we can recover the second order accuracy.

Define the discrete error as

$$e_j^n = u_j^n - U_j^n,$$

where u_j^n represents the exact solution at (x_j, t_n) and U_j^n represents its approximation. The error satisfies the equation

$$\frac{e_j^{n+1} - e_j^n}{\Delta t} = \frac{D}{2(\Delta x)^\alpha}(\delta_{l,2,abs}^\alpha e_j^n + \delta_{l,2,abs}^\alpha e_j^{n+1}) + T_j^n, \tag{5.39}$$

with $e_0 = 0$, $e_N = 0$ and where T_j^n is the local truncation error. We define $e^n = (e_1^n, \ldots, e_{N-1}^n)$, to obtain the vectorial form

$$\frac{e^{n+1} - e^n}{\Delta t} = \frac{D}{(\Delta x)^\alpha}\delta_{l,2,abs}^\alpha \left(\frac{e^n + e^{n+1}}{2}\right) + T^n. \tag{5.40}$$

Since we have proved the numerical method is stable, we can easily prove its convergence for the cases we have consistency and show that the rate of convergence is at least of the same order as the order of accuracy. However, as we have seen, we may have inconsistency for some functions, in particular for some of the cases of interest which are the functions that behave near the left boundary as x^γ. Additionally the numerical experiments point out that the rate of convergence can be higher than the order of accuracy, although never exceeding two, and therefore the result on the convergence should illustrate the improvement of the rate of convergence versus the order of accuracy. To prove such result we need a different approach from the one used in the stability analysis.

The following lemma is an improvement of Lemma 5.5, presented in Sect. 5.5.1. This will be the tool necessary to obtain the convergence result that allow us to infer that for some cases with no consistency we can achieve convergence.

Lemma 5.6 *Let* $\mathbf{e} = (e_1, \ldots, e_{N-1})$ *with* $e_0 = e_N = 0$. *Then*

$$\frac{1}{(\Delta x)^\alpha}(\delta_{l,2,abs}^\alpha \mathbf{e}, \mathbf{e}) \leq -C_\alpha \sum_{j=1}^{N-1} \Delta x e_j^2 a_{j,\alpha},$$

where $a_{j,\alpha} = (x_{j-1})^{-\alpha} + (x_{N-(j+1)})^{-\alpha}$ *and* C_α *is a positive constant.*

Proof We have

$$(\delta_{l,2,abs}^\alpha \mathbf{e}, \mathbf{e}) = \frac{1}{\Gamma(4-\alpha)} \sum_{j=1}^{N-1} \Delta x e_j \left(\sum_{k=-1}^{j} b_k e_{j-k} \right).$$

Noting that we can put together

$$b_{-1} e_j e_{j+1} + b_1 e_{j+1} e_j \quad \text{and} \quad b_{-1} + b_1 \geq 0,$$

by Young's inequality, we can obtain

$$(\delta_{l,2,abs}^\alpha \mathbf{e}, \mathbf{e}) \leq \frac{1}{2\Gamma(4-\alpha)} \sum_{j=1}^{N-1} e_j^2 \left(\sum_{k=-1}^{j-1} b_k \right) + \frac{1}{2\Gamma(4-\alpha)} \sum_{j=1}^{N-1} e_j^2 \left(\sum_{k=-1}^{N-(j+1)} b_k \right).$$

Therefore

$$(\delta_{l,2,abs}^\alpha \mathbf{e}, \mathbf{e}) \leq \frac{1}{2\Gamma(4-\alpha)} \sum_{j=1}^{N-1} e_j^2 \left(\sum_{k=-1}^{j-1} b_k + \sum_{k=-1}^{N-(j+1)} b_k \right).$$

In view of Lemma 5.4, then

$$\frac{1}{(\Delta x)^\alpha} (\delta_{l,2,abs}^\alpha \mathbf{e}, \mathbf{e}) \leq -\frac{C_\alpha}{2\Gamma(4-\alpha)} \sum_{j=1}^{N-1} \Delta x e_j^2 (\Delta x)^{-\alpha} ((j-1)^{-\alpha} + (N-(j+1))^{-\alpha}).$$

We obtain the desired result. $\square$

We can now state the convergence result.

Theorem 5.7 (Convergence) *Let* $e^k := (e_1^k, \ldots, e_{N-1}^k)$ *be a solution of the numerical method (5.39), for a given initial condition* $e^0 := (e_1^0, \ldots, e_{N-1}^0)$ *and the boundary conditions* e_0^k *and* e_N^k *are equal to zero. Let*

$$\|R^n\|^2 = \sum_{j=1}^{N-1} \Delta x (x_j)^\alpha (T_j^n)^2.$$

Then, we have

$$\max_{0 \leq k \leq M} \|e^k\| \leq \left[\|e^0\|^2 + C_\alpha \sum_{n=0}^{M-1} \Delta t \|R^n\|^2 \right]^{1/2},$$

for a positive constant C_α.

Proof Let $e^{n+1/2} = (e^{n+1} + e^n)/2$. By Young's inequality, for all $\epsilon_j > 0$, we have

$$\left(T^n, e^{n+1/2}\right) = \sum_{j=1}^{N-1} \Delta x \, T_j^n e_j^{n+1/2} \leq \sum_{j=1}^{N-1} \Delta x \left(\epsilon_j (T_j^n)^2 + \frac{1}{4\epsilon_j}(e_j^{n+1/2})^2\right).$$

Therefore

$$\frac{1}{2\Delta t}(||e^{n+1}||^2 - ||e^n||^2) \leq \frac{D}{(\Delta x)^\alpha}\left(\delta_{l,2,abs}^\alpha\left(e^{n+1/2}\right), e^{n+1/2}\right)$$
$$+ \sum_{j=1}^{N-1} \Delta x \left(\epsilon_j (T_j^n)^2 + \frac{1}{4\epsilon_j}(e_j^{n+1/2})^2\right).$$

In view of the previous lemma, for $a_{j,\alpha} = (x_{j-1})^{-\alpha} + (x_{N-(j+1)})^{-\alpha}$,

$$\frac{D}{(\Delta x)^\alpha}\left(\delta_{l,2,abs}^\alpha\left(e^{n+1/2}\right), e^{n+1/2}\right) \leq -C_\alpha \sum_{j=1}^{N-1} \Delta x (e_j^{n+1/2})^2 a_{j,\alpha}.$$

Then

$$\frac{1}{2\Delta t}(||e^{n+1}||^2 - ||e^n||^2) \leq -C_\alpha \sum_{j=1}^{N-1} \Delta x (e_j^{n+1/2})^2 a_{j,\alpha}$$
$$+ \sum_{j=1}^{N-1} \Delta x \left(\epsilon_j (T_j^n)^2 + \frac{1}{4\epsilon_j}(e_j^{n+1/2})^2\right)$$

and

$$||e^{n+1}||^2 \leq ||e^n||^2 - C_\alpha 2\Delta t \sum_{j=1}^{N-1} \Delta x (e_j^{n+1/2})^2 a_{j,\alpha}$$
$$+ 2\Delta t \sum_{j=1}^{N-1} \Delta x \left(\epsilon_j (T_j^n)^2 + \frac{1}{4\epsilon_j}(e_j^{n+1/2})^2\right).$$

Let ϵ_j be such that

$$- C_\alpha a_{j,\alpha} + \frac{1}{4\epsilon_j} \leq 0,$$

that is, $\epsilon_j \geq 1/(4C_\alpha a_{j,\alpha})$. Then

$$||e^{n+1}||^2 \leq ||e^n||^2 + 2\Delta t \sum_{j=1}^{N-1} \Delta x \epsilon_j (T_j^n)^2.$$

In particular, if $\epsilon_j = (x_j)^\alpha/4C_\alpha$, then

$$||e^{n+1}||^2 \leq ||e^n||^2 + \frac{1}{2C_\alpha}\Delta t \sum_{j=1}^{N-1} \Delta x (x_j)^\alpha (T_j^n)^2.$$

Upon summation through all $n = 0, 1, \ldots, k$ this implies that

$$||e^k||^2 \leq ||e^0||^2 + C_\alpha \sum_{n=0}^{k-1} \Delta t ||R^n||^2,$$

for all k, $1 \leq k \leq M$, and with

$$||R^n||^2 = \sum_{j=1}^{N-1} \Delta x (x_j)^\alpha (T_j^n)^2.$$

It follows that

$$\max_{0 \leq k \leq M} ||e^k||^2 \leq ||e^0||^2 + C_\alpha \sum_{n=0}^{M-1} \Delta t ||R^n||^2.$$

Hence

$$\max_{0 \leq k \leq M} ||e^k|| \leq \left[||e^0||^2 + C_\alpha \sum_{n=0}^{M-1} \Delta t ||R^n||^2 \right]^{1/2}.$$

$\square$

Let us compute the truncation error and the global error for some particular cases to understand better the previous result.

Suppose that we have a problem defined in the domain $[0, 1] \times [0, 1]$ and that the exact solution of the problem behaves similarly to the following function

$$u(x, t) = g(t)x^\gamma,$$

for a fixed $\gamma > 0$.

Consider the uniform mesh in space and time, that is,

$$x_j = j\Delta x, \quad \text{for} \quad j = 0, \ldots, N; \quad t_n = t_{n-1} + \Delta t, \ n = 0, \ldots, M,$$

with $x_N = 1$ and $t_M = 1$. We will discuss four particular cases, when $\gamma = 3, 2, \alpha, 1$.

For $\gamma = 3$, by Corollary 5.1, the spatial truncation error is dominated by the leading term

$$T_j^L := (\Delta x)^2 (x_j)^{1-\alpha} \left| u^{(3)}(x_0) \right|.$$

Therefore, according to Theorem 5.7, the rate of convergence will be controlled by the upper bound of $\|R^n\|$, dominated by

$$\|R^L\| := \left(\sum_{j=1}^{N-1} \Delta x (x_j)^\alpha (T_j^L)^2 \right)^{1/2}.$$

That is,

$$\|R^L\|^2 = \sum_{j=1}^{N-1} \Delta x (x_j)^\alpha \left((\Delta x)^2 (x_j)^{1-\alpha} \left| u^{(3)}(x_0) \right| \right)^2.$$

We have

$$\|R^L\|^2 := (\Delta x)^4 \left(u^{(3)}(x_0) \right)^2 \sum_{j=1}^{N-1} \Delta x (x_j)^{2-\alpha} \le 36 (\Delta x)^4 (x_N)^{3-\alpha}.$$

Therefore the rate of convergence expected is at least $p = 2$.

For $\gamma = 2$ we can not proceed in the same way as for $\gamma = 3$, since $u^{(2)}(x_0)$ is not zero and the leading term of the truncation error is now

$$T_j^L := (\Delta x)^2 (x_j)^{-\alpha} \left| u^{(2)}(x_0) \right|.$$

Consequently, by Theorem 5.7, we have

$$\|R^L\|^2 := (\Delta x)^4 \left(u^{(2)}(x_0) \right)^2 \sum_{j=1}^{N-1} \Delta x (x_j)^{-\alpha} = 4(\Delta x)^{5-\alpha} \sum_{j=1}^{N-1} j^{-\alpha}$$

$$\le 4(\Delta x)^{5-\alpha} \xi(\alpha),$$

where $\xi(z)$ is the Riemann zeta function. Therefore, the rate of convergence is at least of order $p = 2.5 - \alpha/2$, which for $\alpha \in (1, 2)$ will be between 1.5 and 2.

For $\gamma = 1$, we have that $u^{(1)}(x_0)$ is not zero and the leading term of the truncation error is now

$$T_j^L := (\Delta x)^2 (x_j)^{-1-\alpha} \left| u^{(1)}(x_0) \right|.$$

Similarly, from Theorem 5.7, it follows that

$$\|R^L\|^2 := (\Delta x)^4 \left(u^{(1)}(x_0) \right)^2 \sum_{j=1}^{N-1} \Delta x (x_j)^{-2-\alpha}$$

$$= (\Delta x)^{3-\alpha} \sum_{j=1}^{N-1} j^{-2-\alpha} \le (\Delta x)^{3-\alpha} \xi(\alpha + 2).$$

The rate of convergence is at least of order $p = 1.5 - \alpha/2$, which means will be between 0.5 and 1. Note that for this case we had no consistency when $\alpha = 1.6$ and $\alpha = 1.8$ and now, through the upper bound of Theorem 5.7, we predict convergence.

For $\gamma = \alpha$, we need to turn to Corollary 5.4, since we have $u^{(2)}(x) = \alpha(\alpha - 1)x^{\alpha-2}$ and the second order derivative is not defined at $x_0 = 0$, for $\alpha \in (1, 2)$. In this case, we have $\sup_{x \in (0,1)} x^{2-\alpha} u^{(2)}(x) < \infty$. The truncation error is dominated by (see the quantity (5.24), named A_1 in the proof of Theorem 5.5),

$$T_j^L := (\Delta x)^2 (x_j)^{-\alpha} (\Delta x)^{\alpha-2}.$$

Therefore, we have

$$\|R^L\|^2 := (\Delta x)^{2\alpha} \sum_{j=1}^{N-1} \Delta x (x_j)^{-\alpha} = (\Delta x)^{\alpha+1} \sum_{j=1}^{N-1} j^{-\alpha} \le (\Delta x)^{\alpha+1} \xi(\alpha).$$

The rate of convergence expected is at least $p = \alpha/2 + 0.5$. Note that the order of accuracy associated to this example was only 0.5.

The consistency of a numerical method in the presence of an absorbing boundary condition has been studied and we have seen that the order of accuracy of the local truncation error can be lower than two and for some cases we have inconsistency.

Regarding the convergence it has been shown that for some of the inconsistent cases, we can achieve convergence in the L_2 discrete norm. Moreover, the rate of convergence can be higher than the order of accuracy predicted by the consistency results, although not exceeding the order two.

The left Riemann-Liouville fractional derivative is non-local and its approximation is a quadrature formula involving all the discrete points at the left of the point the derivative is being computed. The weights of the quadrature formula faraway from this point tend to zero very quickly. Therefore at the inner points, if far away enough from the boundary, the second order accuracy of the approximation can be kept. However, any approximation at the points near the boundary is truncated in a relevant way. The number of points in the quadrature formula is small and now we have a significantly different approximation when compared with the approximation at the inner points.

The truncation of the approximation at the nearest points to the boundary leads to a phenomena similar to what we have seen in classical problems when numerical boundary conditions are needed, many times because an high order scheme is being used and it can not be implemented at the points near the boundary due to the lack of enough interpolation points. Hence at the nearest points to the boundary, we end up with a significantly different approximation from the one used at the inner points.

The results of this section can be generalised to problems that include a right Riemann-Liouville fractional derivative with an absorbing boundary on the right hand side of the domain or to problems including simultaneously a combination of the left and right fractional derivatives with absorbing boundaries on both sides of the domain. It can also be studied to understand how to prove the convergence of the numerical methods, when other approximations of the fractional derivative are considered (see Exercise 5.11).

5.6 The Influence of the Boundaries

We illustrate the effect of the boundaries for different values of α. We consider the approximation of the solution of the fractional diffusion equation without the source term

$$\frac{\partial u(x,t)}{\partial t} = D \frac{\partial^{\alpha}_{abs} u}{\partial x^{\alpha}}(x,t), \tag{5.41}$$

with an initial condition that is an approximation of the Dirac delta function

$$u_0(x) = \frac{1}{\epsilon\sqrt{\pi}} e^{-(x-x_0)^2/\epsilon^2},$$

for a small $\epsilon > 0$. For all figures we have taken $D = 1$, $\epsilon = 0.1$, $x_0 = 1$. In this way we obtain approximations of the fundamental solutions of the equations.

In the next figures we consider two values of α. In Fig. 5.1, we display the solution for $\alpha = 1.3$ and in Fig. 5.2, for $\alpha = 1.8$. We show the solution as we evolve in time, to observe the effect of the boundary on the solution. We illustrate also the problem defined in the open domain, where the left Riemann-Liouville fractional derivative has the initial point of the integral at $-\infty$. By the figures we can observe that in the open domain we have an asymmetric case since we are only considering the left Riemann Liouville fractional derivative. Therefore, the wave has a significant heavy tail on the right hand side of the domain.

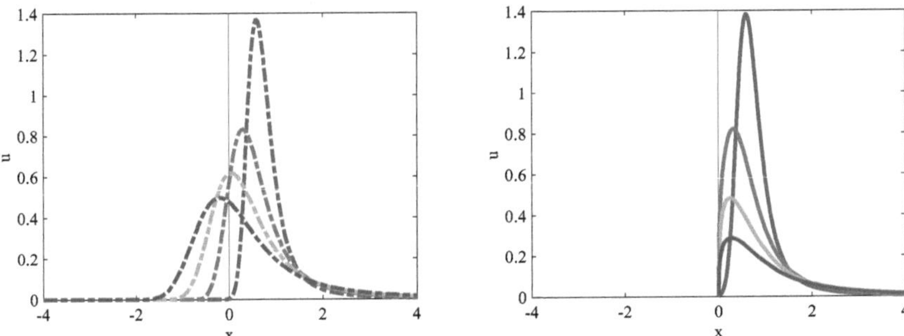

Fig. 5.1 Plots of $u(x, t)$ for $x_0 = 1$, $D = 1$, $\alpha = 1.3$. Open domain on the left $(--)$; Absorbing boundary on the right $(-)$. Evolution in time described for $t = 0.25, 0.5, 0.75, 1$

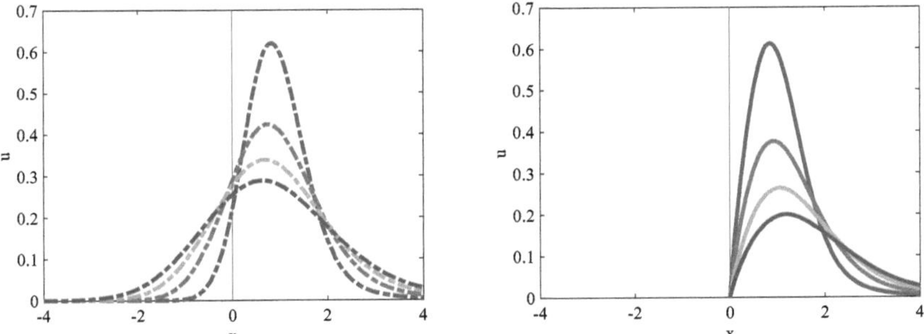

Fig. 5.2 Plots of $u(x, t)$ for $x_0 = 1$, $D = 1$, $\alpha = 1.8$. Open domain on the left $(--)$; Absorbing boundary on the right $(-)$. Evolution in time described for $t = 0.25, 0.5, 0.75, 1$

Near the boundary, the graphical behaviour of the solution shows that its derivative may not exist for the value $x = 0$. In particular for smaller values of α this is quite evident. The steady state solution of the fractional diffusion equation is the combination of the functions $x^{\alpha-1}$ and $x^{\alpha-2}$. The solution that is zero at $x = 0$ is $x^{\alpha-1}$. This is also the function that better describes the behaviour we observe in the previous figures, near the boundary, where the solution is zero and the tangent at this point has inifinite slope.

5.7 Exercises and Solutions

Exercises

5.1 For $\alpha \in (1, 2)$, consider the equation $D_{0+}^{\alpha} u(x) = 0$.

(a) Show that the general solution is a combination of the functions $x^{\alpha-2}$ and $x^{\alpha-1}$.

(b) If we assume the condition $u(0) = 0$, which solution is the appropriate one?

(c) And what happens if we have the boundary condition $D_{0+}^{\alpha-1}u(0) = 0$?

5.2 Consider the transport equation that moves from right to left, that is,

$$\frac{\partial u}{\partial t}(x, t) - V\frac{\partial u}{\partial x}(x, t) = 0, \quad V > 0, \quad x \in (0, 1), \quad t > 0,$$

with $u(1, t) = 0$. There is no boundary condition given at $x = 0$.

(a) Derive the numerical method that consists on the Euler explicit scheme and the centered difference operator for approximating the first order spatial derivative.

(b) Conclude that this approach breaks down at $x = 0$, since no values are defined for $x < 0$.

(c) Use a one-sided operator at the boundary which yield one-order lower accuracy than the one used at inner points.

(d) Prove the numerical method has at least one order accuracy of convergence.

5.3 Consider the coefficient g_k^{α} defined by

$$g_k^{\alpha} = (-1)^k \binom{\alpha}{k}.$$

Prove that

$$\sum_{k=0}^{j} g_{j-k}^{\alpha}k^{\gamma} = \frac{\Gamma(\gamma + 1)}{\Gamma(\gamma + 1 - \alpha)}j^{\gamma-\alpha} - \frac{\alpha}{2}\frac{\Gamma(\gamma + 1)}{\Gamma(\gamma - \alpha)}j^{\gamma-\alpha-1} + O(j^{\gamma-\alpha-2}).$$

5.4 Justify that we have

$$\left(1 + \frac{p}{j}\right)^{\gamma} = 1 + \gamma\frac{p}{j} + O(j^{-2}), \quad j \geq |p|.$$

5.5 Prove that

$$(x_{j+1} - \xi)^{1-\alpha} - (x_j - \xi)^{1-\alpha} \leq \Delta x(x_j - x_0)^{-\alpha}.$$

5.6 Let $u(x, \cdot) \in C^3(0, \infty)$, $u(\cdot, t) \in AC([a, b]) \cap C^4(a, b]$ and u near the left boundary behaves as $(x - a)^{\gamma}$, $\gamma > 0$, that is,

$$\sup_{x\in(a,b)} (x - a)^{n-\gamma}\left|\frac{\partial^n u}{\partial x^n}(x, t)\right| < \infty, \quad n = 2, 3, 4.$$

Prove that the local truncation error of the numerical method (5.31) is given by

$$T_j^n = \frac{(\Delta t)^2}{2} \frac{\partial^3 u(x_j, \tau_n)}{\partial t^3} - \frac{1}{2} \left(\epsilon_T(\Delta x, \alpha, a, x_j, t_{n+1}) + \epsilon_T(\Delta x, \alpha, a, x_j, t_n) \right),$$

for a τ_n such that $t_n \leq \tau_n \leq t_{n+1}$ and ϵ_T denotes the truncation error (5.32).

5.7 Prove that a symmetric strictly diagonally dominant real matrix with nonnegative diagonal entries is strictly positive definite.

5.8 Prove that a real square matrix A is strictly definite positive if and only if $H = (A + A^T)/2$ is strictly definite positive.

5.9 Let $g_k^\alpha \; k = 0, 1, 2, \ldots$ defined by

$$g_k^\alpha = (-1)^k \binom{\alpha}{k}$$

and

$$A = \begin{bmatrix} g_1^\alpha & g_0^\alpha & 0 & \cdots & & 0 \\ g_2^\alpha & & & & & \\ & & \ddots & & & \vdots \\ \vdots & & & \ddots & & 0 \\ & & & & & g_0^\alpha \\ g_{N-1}^\alpha & & \cdots & & g_2^\alpha & g_1^\alpha \end{bmatrix}.$$

(a) Prove that the matrix A is strictly diagonal dominant.

(b) The matrix $-A$ is strictly definite positive.

5.10 Consider the induced 2-norm. Let A be definite positive and $\theta \geq 0$. We have the following properties:

(a) $\|(I + \theta A)^{-1}\| \leq 1$.
(b) $(I + \theta A)^{-1}$ and $I - \theta A$ commute.
(c) $\|(I + \theta A)^{-1}(I - \theta A)\| \leq 1$.

5.11 Consider a fractional diffusion problem, with an absorbing boundary condition at $x = 0$, where the fractional operator is the combination of the left and right Riemann-Liouville fractional derivatives. The domain is $(0, \infty)$ and we assume that u goes to zero, when x goes to infinity. Define the computational domain and show the matricial form of a numerical method.

5.12 Consider the fractional diffusion problem with only the left Riemann-Liouville fractional derivative define in $[a, b]$. Assume we have Dirichlet boundary conditions. Consider the Crank-Nicolson numerical method for the time discretisation and assume that the fractional derivative is approximated with the Grünwald-Letnikov approximation.

(a) Prove the inequality similar to Lemma 5.6, that is,

$$\frac{1}{(\Delta x)^\alpha}(\delta^\alpha_{l,1,abs}e, e) \le -C_\alpha \sum_{j=1}^{N-1} \Delta x e_j^2 a_{j,\alpha}.$$

(b) Prove the stability of the numerical method with the energy method.
(c) Prove the convergence of the numerical method with the energy method.

Solutions

5.1 (a) We can arrive at the solutions $x^{\alpha-2}$ and $x^{\alpha-1}$ by direct calculations. Start
    by noting that

$$I_{0+}^{2-\alpha}u(x) = ax + b,$$

    for a and b real constants.

(b) If we assume the condition $u(0) = 0$, this will force the solution to behave
    according to $x^{\alpha-1}$.

(c) If we have the boundary condition $D_{0+}^{\alpha-1}u(0) = 0$, the solution will
    behave according to $x^{\alpha-2}$. Just compute $D_{0+}^{\alpha-1}(x^{\alpha-2})$ and see what you
    obtain.

5.2 (a) The numerical method is given by

$$U_j^{n+1} = U_j^n + \frac{v}{2}(U_{j+1}^n - U_{j-1}^n), \quad v = V\frac{\Delta t}{\Delta x}.$$

(b) At $j = 0$ we have

$$U_0^{n+1} = U_0^n + \frac{v}{2}(U_1^n - U_{-1}^n)$$

    and we do not have the points U_{-1}^n.

(c) We can use the numerical method

$$U_0^{n+1} = U_0^n + v(U_1^n - U_0^n).$$

(d) Use the energy method similarly to what has been done in Chap. 3.

5.3 See proof of Lemma 3.5 in [89].

5.4 Do the expansion of the binomial series

$$\left(1 + \frac{p}{j}\right)^\gamma = \sum_{k=0}^{\infty} \binom{\gamma}{k}\left(\frac{p}{j}\right)^k \quad \text{with} \quad \left|\frac{p}{j}\right| \le 1.$$

5.5 By noting the equality

$$(x_{j+1} - \xi)^{1-\alpha} - (x_j - \xi)^{1-\alpha}$$

$$= (x_j - x_0)^{-\alpha} \left[\left(\frac{x_{j+1} - \xi}{x_j - x_0} \right)^{-\alpha} (x_{j+1} - \xi) - \left(\frac{x_j - \xi}{x_j - x_0} \right)^{-\alpha} (x_j - \xi) \right]$$

and that, for $\xi \in (x_0, x_1)$, we have

$$\left(\frac{x_j - x_0}{x_{j+1} - \xi} \right)^{\alpha} \leq \left(\frac{x_j - x_0}{x_j - \xi} \right)^{\alpha} \leq 1 \quad \text{and} \quad \left(\frac{x_j - x_0}{x_j - \xi} \right)^{\alpha} \geq 1.$$

It follows that

$$(x_{j+1} - \xi)^{1-\alpha} - (x_j - \xi)^{1-\alpha} \leq (x_j - x_0)^{-\alpha} \left[(x_{j+1} - \xi) - (x_j - \xi) \right]$$

$$\leq \Delta x (x_j - x_0)^{-\alpha}.$$

5.6 The proof can be done following the usual procedure for the Crank-Nicolson method and using the truncation error obtained previously for the fractional operator.

5.7 If a matrix is strictly diagonally dominant and all its diagonal elements are positive, then the real parts of its eigenvalues are positive. This result follows from the Gershgorin circle theorem.

5.8 If A is definite positive, A^T is definite positive and therefore $H = (A + A^T)/2$ is definite positive. On the other hand if $H = (A + A^T)/2$ is definite positive then $x^T A x + x^T A^T x = x^T A x + (x^T A x)^T > 0$.

5.9 (a) The matrix A is strictly diagonal dominant. This comes straightly from the fact that

$$|g_1^{\alpha}| > \sum_{j=0, j \neq 1}^{N} |g_j^{\alpha}|.$$

(b) We prove that the matrix $-A$ is strictly definite positive by showing that $H = -(A^T + A)/2$ is strictly definite positive. And this is shown by proving that $(A^T + A)/2$ is a symmetric matrix and strictly diagonal dominant. We have that

$$H_{i,j} = \begin{cases} -g_{j-k+1}^{\alpha} & \text{for } k < j - 1 \\ -(g_0^{\alpha} + g_2^{\alpha}) & \text{for } k = j - 1 \\ -2g_1^{\alpha} & \text{for } k = j \\ -(g_0^{\alpha} + g_2^{\alpha}) & \text{for } k = j + 1 \\ -g_{j-k+1}^{\alpha} & \text{for } k > j + 1. \end{cases}$$

Obviously H is symmetric and the diagonal entries $-2g_1^\alpha$ are nonnegative. The matrix H is also strictly diagonal dominant since

$$|2g_1^\alpha| > \sum_{j=0, j\neq 1}^{N} 2|g_j^\alpha|.$$

Hence,

$$|H_{i,i}| > \sum_{j=1, j\neq i}^{N} |H_{i,j}|.$$

5.10 (a) First, we show that the inverse of $(I + \theta A)$ exists. We can do that by contradiction. If we assume that the matrix is not invertible, then there exists an x different from the zero vector such that $(I + \theta A)x = 0$. Then, we can conclude that A is not definite positive.

Let v be a general vector. We have that

$$v^T (I + \theta A)^T (I + \theta A)v \geq v^T v. \tag{5.42}$$

This can be easily proved by checking that

$$v^T (I + \theta A)^T (I + \theta A)v = v^T [I + \theta(A + A^T) + \theta^2 A^T A]v.$$

Then, since $A + A^T$ is definite positive because A is definite positive, we have

$$v^T v + \theta v^T (A + A^T)v + \theta^2 v^T \|Av\|^2 \geq v^T v.$$

Now taking a general w, let $v = (I + \theta A)^{-1}w$ and substitute it in (5.42),

$$w^T (I + \theta A)^{-T}(I + \theta A)^T (I + \theta A)(I + \theta A)^{-1}w$$
$$\geq w^T (I + \theta A)^{-T}(I + \theta A)^{-1}w.$$

Hence,

$$w^T w \geq \|(I + \theta A)^{-1}w\|^2$$

Therefore, for all w, we have

$$\frac{\|(I + \theta A)^{-1}w\|_2^2}{\|w\|^2} \leq 1.$$

This means that

$$\|(I + \theta A)^{-1}\| \leq 1.$$

(b) We want to prove that $(I + \theta A)^{-1}$ and $I - \theta A$ commute. We can prove this directly by doing some algebraic manipulations. We obtain

$$
\begin{aligned}
(I + \theta A)^{-1}(I - \theta A) &= (I + \theta A)^{-1}(I - \theta A)[(I + \theta A)(I + \theta A)^{-1}] \\
&= (I + \theta A)^{-1}[(I - \theta A)[(I + \theta A)](I + \theta A)^{-1} \\
&= (I + \theta A)^{-1}[(I + \theta A)[(I - \theta A)](I + \theta A)^{-1} \\
&= [(I + \theta A)^{-1}(I + \theta A)](I - \theta A)(I + \theta A)^{-1} \\
&= (I - \theta A)(I + \theta A)^{-1}.
\end{aligned}
$$

(c) We want to prove that $\|(I + \theta A)^{-1}(I - \theta A)\| \leq 1$.
    We start to notice that, for any vector v,

$$v^T(I - \theta A)^T(I - \theta A)v \leq v^T(I + \theta A)^T(I + \theta A)v. \qquad (5.43)$$

This inequality can be easily checked by noticing that $A + A^T$ is definite positive and on the left hand side of the previous inequality we have

$$v^T v - \theta v^T(A + A^T)v + \theta^2\|Av\|^2$$

and on the right hand side

$$v^T v + \theta v^T(A + A^T)v + \theta^2\|Av\|^2.$$

Now taking a general w, let $v = (I + \theta A)^{-1}w$ and substitute it in (5.43). We obtain

$$
\begin{aligned}
w^T(I + \theta A)^{-T}(I - \theta A)^T(I - \theta A)(I + \theta A)^{-1}w \\
\leq w^T(I + \theta A)^{-T}(I + \theta A)^T(I + \theta A)(I + \theta A)^{-1}w.
\end{aligned}
$$

Hence,

$$w^T((I - \theta A)(I + \theta A)^{-1})^T(I - \theta A)(I + \theta A)^{-1}w \leq w^T w.$$

This means that, for all $w \neq 0$,

$$\|(I - \theta A)(I + \theta A)^{-1}w\|^2 \leq \|w\|^2$$

and consequently

$$||(I - \theta A)(I + \theta A)^{-1}|| = \sup_{w \neq 0} \frac{||(I - \theta A)(I + \theta A)^{-1}w||}{||w||} \leq 1.$$

Using (b) we can conclude that

$$||(I + \theta A)^{-1}(I - \theta A)|| = ||(I - \theta A)(I + \theta A)^{-1}|| \leq 1.$$

This proof can be seen in some works such as [172], in the context of the matricial form of a numerical method.

5.11 Similar to what we have done in Sect. 5.4.

5.12 (a) See the proof of Lemma 2 in [138].

(b) To prove the stability of the numerical method with the energy method just proceed similarly to Theorem 5.6, where the approximation based on the linear spline approximation of the solution has been considered.

(c) To prove the convergence of the numerical method, with the energy method, just proceed similarly to Theorem 5.7, where once more the approximation based on the linear spline approximation of the solution has been considered.

Chapter 6
Fractional Differential Equations in Bounded Domains: Reflecting Boundaries

In this chapter, we describe how the fractional diffusion equation can take in consideration a reflecting boundary. The discussion on reflecting boundaries has been more controversial than the study on absorbing boundaries. There are several ways to define reflecting boundaries and they lead to substantially different solutions. Due to the non-locality of these problems, the inclusion of a boundary condition results in modifying the large-scale model. We start with a review about what has been considered until the moment in the context of Lévy flights. This is a very active area of research with still many open questions, which means we still do not understand this fully. Nevertheless, the subject already deserves to be included in an advanced course. After introducing two models, the numerical methods for both models are described. The properties of the numerical methods are discussed using tools provided in previous chapters.

6.1 Overview on Reflecting Boundary Conditions

Anomalous diffusive transport, in particular superdiffusion, arises in a large variety of physical problems. It frequently happens that we have to apply boundary conditions when considering experimental devices and attempting to check a model for mass transport in a given medium. Due to non-locality, it is not obvious how to incorporate a boundary condition in a scenario based on Lévy flights, and significant modifications of the macroscopic model may be expected. In fact, the presence of boundaries may modify the nonlocal spatial operator. In literature, when discussing Lévy flights in the one dimensional half-space, the boundary conditions mainly considered have been absorbing or reflecting boundaries.

Fractional diffusion equations, related to Lévy flights, with boundaries have been studied in a considerable number of works, essentially from two separate perspectives, which are related to the mathematical or the physical point of views.

E. Sousa, *Finite Difference Methods for Fractional Diffusion Equations*, Lecture Notes in Mathematics 2389, https://doi.org/10.1007/978-3-032-11222-4_6

Absorbing boundary conditions have been imposed by assuming zero outside the problem domain as we have seen in the previous chapter. Regarding reflecting boundary conditions several formulations have been proposed [6, 22, 37, 44, 48, 49, 74, 77, 78].

We present two approaches that appeared in literature several years ago within the physical context. More recently, they have been under scrutiny in the context of deriving numerical methods for these type of problems [8, 77, 78]. In [8] the reflecting boundary condition for fractional diffusion equations is derived using a mass balance approach. This boundary is defined in terms of a Riemann-Liouville fractional derivative. It is also shown that a reflecting boundary condition defined in terms of the Caputo fractional derivative is unsuitable for modeling the boundary, since the resulting boundary value problem is not positivity preserving. In [77, 78] another reflecting boundary condition is presented and it is shown how it modifies the space-time fractional evolution equation, by changing the kernel of the non-local fractional operator.

Some other approaches have been appearing in literature. For instance, in [44] the reflecting boundary is given in terms of a nonlocal normal derivative operator similar to the fractional derivative and a probabilistic interpretation of the problem is presented. However, we will not be discussing it here, since this approach has been less explored from the point of view of deriving finite difference methods.

The numerical methods presented in this chapter follows the ideas discussed in the previous chapters. In this way we also illustrate how a numerical method changes for a problem under the influence of a boundary condition.

We start to revisit the problem without boundaries and then we describe the two resulting problems after the influence of the two different boundaries. Subsequently, we consider numerical methods based in the Grünwald-Letnikov approximation and discuss the matrix iteration of the numerical methods, highlighting the differences between having boundaries or not. In another section, we consider a numerical method based in the Riemann-Liouville integral approximation and discuss consistency and stability. This will be followed by the illustration of the solutions of the problems with both types of reflecting boundaries.

6.2 Problem Formulation

In previous chapters, we came across the formulation of a fractional differential equation defined in the open domain. Subsequently, in the last chapter we have discussed what happens if we include an absorbing boundary condition. In the next section, we reexamine the problem formulation in the open domain and then explore the impact associated with the presence of a reflecting boundary, specifically considering two distinct types on one side of the domain.

6.2.1 Open Domain Revisited

The superdiffusive model associated with Lévy flights is defined in the whole real line and the governing equation involves Riemann-Liouville fractional derivatives [98]. The left and right Riemann-Liouville fractional derivatives of order α, $\alpha \in (1, 2)$, for $x \in \mathbb{R}$, are given respectively by

$$\frac{\partial^\alpha u}{\partial x^\alpha}(x, t) = \frac{1}{\Gamma(2 - \alpha)} \frac{\partial^2}{\partial x^2} \int_{-\infty}^{x} u(\xi, t)(x - \xi)^{1-\alpha} d\xi, \tag{6.1}$$

$$\frac{\partial^\alpha u}{\partial(-x)^\alpha}(x, t) = \frac{1}{\Gamma(2 - \alpha)} \frac{\partial^2}{\partial x^2} \int_{x}^{\infty} u(\xi, t)(\xi - x)^{1-\alpha} d\xi. \tag{6.2}$$

The fractional differential equation describing the superdiffusive model in the open domain, for $\alpha \in (1, 2)$ and $\beta \in (-1, 1)$, can be stated as

$$\frac{\partial u(x, t)}{\partial t} = D \left(\frac{1 + \beta}{2} \frac{\partial^\alpha u}{\partial x^\alpha}(x, t) + \frac{1 - \beta}{2} \frac{\partial^\alpha u}{\partial(-x)^\alpha}(x, t) \right) + f(x, t), \tag{6.3}$$

where D is the diffusive parameter and $f(x, t)$ is a source term. The parameter α describes the tail of the solution and the parameter β is the skewness and specifies if the solution is skewed to the left ($\beta < 0$), right ($\beta > 0$) of if it is symmetric ($\beta = 0$).

6.2.2 The Symmetric Boundary Wall

A superdiffusive problem with a left reflecting wall has been formulated in [77, 78]. The model consists of a reflecting wall restraining the diffusing particles to a semi-infinite domain. That barrier can be viewed as a force field applied to the particles. It is assumed that the particles arriving at the boundary are bounced back as in elastic collisions, that is, if they reach the position $x = -a$ with $a > 0$, then they will end at $x = a$, describing the mirror trajectory with respect to the wall. In the context of a porous medium, such a boundary may represent a wall permeable to the fluid, but impermeable to the tracer.

Let $u(x, t)$ be the solution on $x > 0$ and $u^*(x, t)$ be the even extension of u, defined by

$$u^*(x, t) = u(x, t), \quad x > 0, \tag{6.4}$$

$$u^*(x, t) = u(-x, t), \quad x < 0. \tag{6.5}$$

See illustration of the reflecting wall in Fig. 6.1.

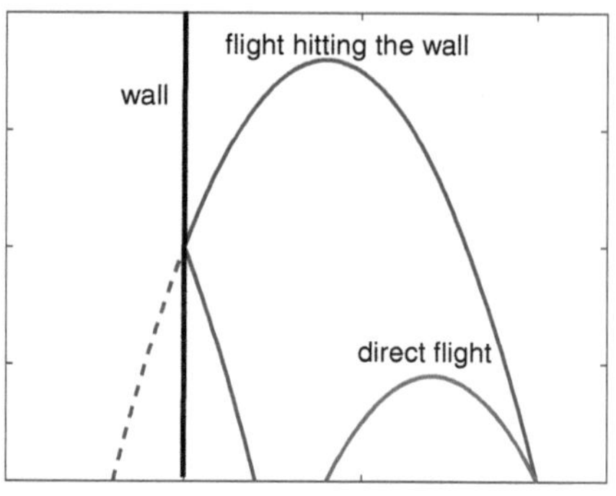

Fig. 6.1 Ilustration of the reflecting boundary according to (6.4) and (6.5)

The function u^* satisfies, in $\mathbb{R}$, the equation

$$\frac{\partial u^*}{\partial t}(x,t) = D\left(\frac{1+\beta}{2}\frac{\partial^\alpha u^*}{\partial x^\alpha}(x,t) + \frac{1-\beta}{2}\frac{\partial^\alpha u^*}{\partial(-x)^\alpha}(x,t)\right) + f(x,t). \tag{6.6}$$

Mathematically, we can write this problem defined in $x > 0$ by incorporating the wall condition, $u(x,t) = u(-x,t), x < 0$, in the definition of the fractional operator. Taking in consideration that $u(x,t) = u(-x,t)$, for $x < 0$, the left Riemann-Liouville fractional derivative is affected by this condition and we have, for $x > 0$,

$$\frac{\partial^\alpha u}{\partial x^\alpha}(x,t) = \frac{1}{\Gamma(2-\alpha)}\frac{\partial^2}{\partial x^2}\int_{-\infty}^0 u(\xi,t)(x-\xi)^{1-\alpha}d\xi$$

$$+\frac{1}{\Gamma(2-\alpha)}\frac{\partial^2}{\partial x^2}\int_0^x u(\xi,t)(x-\xi)^{1-\alpha}d\xi$$

$$=\frac{1}{\Gamma(2-\alpha)}\frac{\partial^2}{\partial x^2}\int_{-\infty}^0 u(-\xi,t)(x-\xi)^{1-\alpha}d\xi$$

$$+\frac{1}{\Gamma(2-\alpha)}\frac{\partial^2}{\partial x^2}\int_0^x u(\xi,t)(x-\xi)^{1-\alpha}d\xi.$$

By doing a change of variables we obtain what we will define as the reflecting left Riemann-Liouville fractional derivative, for $x > 0$,

$$\frac{\partial^\alpha_{ref} u}{\partial x^\alpha}(x,t) := \frac{1}{\Gamma(2-\alpha)}\frac{\partial^2}{\partial x^2}\int_0^\infty u(\xi,t)(x+\xi)^{1-\alpha}d\xi$$

$$+\frac{1}{\Gamma(2-\alpha)}\frac{\partial^2}{\partial x^2}\int_0^x u(\xi,t)(x-\xi)^{1-\alpha}d\xi. \tag{6.7}$$

The right Riemann-Liouville fractional derivative is not affected by the reflecting wall. Consequently, it continues to be defined by (6.2), for $x > 0$.

Formally, when subjected to a reflecting wall we have the following problem

$$\frac{\partial u}{\partial t}(x,t) = D\left(\frac{1+\beta}{2}\frac{\partial^\alpha_{ref} u}{\partial x^\alpha}(x,t) + \frac{1-\beta}{2}\frac{\partial^\alpha u}{\partial(-x)^\alpha}(x,t)\right) + f(x,t), \quad x > 0,$$
(6.8)

subjected to the conditions

$$u(x,t) = u(-x,t), \quad \text{for all} \quad x < 0,$$
(6.9)

and with an initial condition

$$u(x,0) = u_0(x), \ x \geq 0.$$

6.2.3 The Fractional Neumann Boundary

Developing accurate and physically meaningful reflecting boundary conditions for the fractional diffusion equation requires careful considerations. Suppose that now the mass that was removed from the system in the model of the previous chapter will instead be preserved, and moved to the boundary. Unlike the traditional diffusion setup, this mass can come from far inside the domain and not just adjacent points. To describe this approach [6, 8], a fractional Neumann boundary condition has been considered. This boundary condition can also be found in [60, 144, 155].

Define, for $n - 1 < \alpha < n$, the left Riemann-Liouville fractional derivative, starting at $x = 0$,

$$\frac{\partial^\alpha_0 u}{\partial x^\alpha}(x,t) = \frac{1}{\Gamma(n-\alpha)}\frac{\partial^n}{\partial x^n}\int_0^x u(\xi,t)(x-\xi)^{n-1-\alpha}d\xi.$$
(6.10)

Formally, for $\alpha \in (1,2)$, we have the problem, defined by the equation

$$\frac{\partial u}{\partial t}(x,t) = D\left(\frac{1+\beta}{2}\frac{\partial^\alpha_0 u}{\partial x^\alpha}(x,t) + \frac{1-\beta}{2}\frac{\partial^\alpha u}{\partial(-x)^\alpha}(x,t)\right) + f(x,t), \ x > 0,$$
(6.11)

with the boundary condition

$$\frac{\partial^{\alpha-1}_0 u}{\partial x^{\alpha-1}}(0,t) = 0, \ t > 0,$$
(6.12)

and the initial condition

$$u(x, 0) = u_0(x), \quad x \geq 0. \tag{6.13}$$

When $\alpha = 2$, Eq. (6.12) is the well-known Neumann boundary condition

$$\frac{\partial u}{\partial x}(0, t) = 0.$$

6.3 Matricial Form and the Iterative Matrix

In this section, we present the numerical methods for the two previous problems and describe its matricial form. We consider the fractional diffusion problem (6.11), for the simpler case $\beta = 1$ and without source term. We approximate the left Riemann-Liouville fractional derivative by the Grünwald-Letnikov approximation discussed in the fourth chapter. The main goal of this section is to show how the presence of boundary conditions changes the structure of the iterative matrix.

6.3.1 Open Domain Revisited

For the problem defined in the whole real line, the domain discretisation is given by

$$\{(x_j, t_n) : x_j = j\Delta x, \ j \in \mathbb{Z}; \ t_n = n\Delta t, \ n \in \mathbb{N}_0\}.$$

We approximate the left Riemann-Liouville fractional derivative by the Grünwald-Letnikov approximation. Define the Grünwald-Letnikov coefficients, for all $\alpha > 0$, using the following recurrence formula

$$g_0^\alpha = 1, \quad g_{k+1}^\alpha = -\frac{\alpha - k}{k + 1} g_k^\alpha, \quad k \geq 0. \tag{6.14}$$

The Grünwald-Letnikov approximation, at (x_j, t_n), as studied in the third chapter, is given by

$$\frac{\partial^\alpha u}{\partial x^\alpha}(x_j, t_n) \approx \frac{1}{(\Delta x)^\alpha} \sum_{k=0}^{\infty} g_k^\alpha u(x_{j-k+1}, t_n). \tag{6.15}$$

Let U_j^n represent the approximate solution of $u(x_j, t_n)$ in the discrete domain and define

$$\mu_\alpha = \frac{D\Delta t}{(\Delta x)^\alpha}.$$

The Euler explicit numerical method to approximate the fractional diffusion equation is given by

$$U_j^{n+1} = U_j^n + \mu_\alpha \sum_{k=0}^{\infty} g_k^\alpha U_{j-k+1}^n, \quad \text{for all } j \in \mathbb{Z}. \tag{6.16}$$

The matricial form of the numerical method in the open domain takes in consideration that the function goes to zero as we go to infinity and we have

$$\mathbf{U}^{n+1} = (\mathbf{I} + \mu_\alpha \mathbf{A})\mathbf{U}^n,$$

with $\mathbf{U}^n = [U_{-N}^n, \ldots, U_N^n]^T$, $\mathbf{I}$ is the identity matrix and the matrix $\mathbf{A}$ is given by

$$\mathbf{A} = \begin{bmatrix} g_1^\alpha & g_0^\alpha & 0 & \cdots & 0 & 0 \\ g_2^\alpha & g_1^\alpha & g_0^\alpha & \cdots & 0 & 0 \\ g_3^\alpha & g_2^\alpha & g_1^\alpha & \cdots & 0 & 0 \\ \vdots & \vdots & \vdots & & \vdots & \vdots \\ g_{2N+1}^\alpha & g_{2N}^\alpha & g_{2N-1}^\alpha & \cdots & g_2^\alpha & g_1^\alpha \end{bmatrix}.$$

In the fourth chapter we have studied the accuracy of the approximation (6.15) and we have seen that under certain regularity conditions of the solution u, this approximation has an order of accuracy one. Since we have considered the Euler explicit numerical method, we know this will lead to a full scheme that is first order accurate.

Some numerical linear algebra techniques have been developed and suggested to improve the performance of the above linear system, see for instance [9]. We will not discuss those investigations in this book. Regarding the stability of the numerical method, since we are in the open domain we can use the von Neumann analysis or Fourier analysis as also discussed previously.

We give here the result, presented in the third chapter, on the stability,

Theorem 6.1 *If the numerical method (6.16) is von Neumann stable, then* $\mu_\alpha \leq 2^{1-\alpha}$.

Proof See the proof of Theorem 4.9 in Chap. 4. $\qquad\qquad\square$

For linear problems with a boundary, the von Neumann stability analysis, in general, gives a necessary condition for the stability of the numerical method.

6.3.2   The Symmetric Boundary Wall

When we have a reflecting boundary condition at $x = 0$, the left fractional derivative is modified to (6.7). In a similar way its approximation also changes. Because

$$U_{-i+1}^n = U_{-j+i-1}^n,$$

the approximation, for $j = 0, 1, \ldots, N-1$, becomes

$$\frac{\delta^\alpha_{ref} u(x_j, t)}{(\Delta x)^\alpha} \approx \frac{1}{(\Delta x)^\alpha} \sum_{i=0}^{j+1} g_i^\alpha U_{j-i+1}^n + \frac{1}{(\Delta x)^\alpha} \sum_{i=j+2}^{\infty} g_i^\alpha U_{j-i+1}^n$$

$$= \frac{1}{(\Delta x)^\alpha} \sum_{i=0}^{j+1} g_i^\alpha U_{j-i+1}^n + \frac{1}{(\Delta x)^\alpha} \sum_{i=j+2}^{\infty} g_i^\alpha U_{i-j-1}^n.$$

Consider the explicit Euler scheme to approximate equation (6.8), when $\beta = 1$, given by

$$U_j^{n+1} = U_j^n + \mu_\alpha \delta^\alpha_{ref} U_j^n. \tag{6.17}$$

In this case the matricial form of the problem is

$$\mathbf{U}^{n+1} = (\mathbf{I} + \mu_\alpha \mathbf{A_{Sym}})\mathbf{U}^n,$$

with $\mathbf{U}^n = [U_0^n, \ldots, U_{N-1}^n]^T$, $\mathbf{I}$ is the identity matrix and the matrix $\mathbf{A_{Sym}}$ is

$$\mathbf{A_{Sym}} = \begin{bmatrix} g_1^\alpha & g_0^\alpha + g_2^\alpha & & g_3^\alpha & \cdots & g_{N-1}^\alpha & g_N^\alpha \\ g_2^\alpha & g_1^\alpha + g_3^\alpha & g_0^\alpha + g_4^\alpha & \cdots & g_N^\alpha & g_{N+1}^\alpha \\ g_3^\alpha & g_2^\alpha + g_4^\alpha & g_1^\alpha + g_5^\alpha & \cdots & g_{N+1}^\alpha & g_{N+2}^\alpha \\ \vdots & \vdots & \vdots & & \vdots & \vdots \\ g_N^\alpha & g_{N-1}^\alpha + g_{N+1}^\alpha & g_{N-2}^\alpha + g_{N+2}^\alpha & \cdots & g_2^\alpha + g_{2N-2}^\alpha & g_1^\alpha + g_{2N-1}^\alpha \end{bmatrix}.$$

Highlighted in gray are the values added to the matrix $\mathbf{A}$ because of the presence of this reflecting boundary. This problem is equivalent to a problem defined in the real line and therefore the stability conditions of the numerical method are expected to be similar to those obtained for the open problem.

6.3.3 The Fractional Neumann Boundary

Suppose we are in a bounded domain and that the approximation (6.15) is considered. The sum in (6.15) will include all the points until the boundary and not beyond the boundary. We arrive at the following numerical method to approximate the fractional differential equation

$$U_j^{n+1} = U_j^n + \mu_\alpha \sum_{i=0}^{j+1} g_i^\alpha U_{j-i+1}^n, \quad j = 0, \ldots, N-1. \tag{6.18}$$

However, in this equation, we do not see any information regarding the boundary. To enforce the boundary condition (6.12), we can modify the numerical method (6.18) at the boundary point $j = 0$. Taking in consideration the properties of the Riemann-Liouville fractional derivative, see Theorem 1.11 in Chap. 1, the differential equation (6.11), at $x = 0$, can be written as

$$\frac{\partial u}{\partial t}(0, t) = D \frac{\partial}{\partial x} \left(\frac{\partial_0^{\alpha-1} u}{\partial x^{\alpha-1}} \right)(0, t).$$

We can use the Euler approximation for the time derivative, that is,

$$\frac{\partial u}{\partial t}(0, t) \approx \frac{U_0^{n+1} - U_0^n}{\Delta t}.$$

A first order approximation for the first order spatial derivative allow us to write

$$\frac{\partial}{\partial x} \left(\frac{\partial_0^{\alpha-1} u}{\partial x^{\alpha-1}} \right)(0, t) \approx \frac{1}{\Delta x} \left(\frac{\partial_0^{\alpha-1} u}{\partial x^{\alpha-1}}(x_1, t) - \frac{\partial_0^{\alpha-1} u}{\partial x^{\alpha-1}}(0, t) \right). \tag{6.19}$$

We know the value of the second term on the right hand side of (6.19), since this is the given boundary condition. Additionally, we can approximate the fractional derivative at x_1 using the Grünwald-Letnikov approximation, that is,

$$\frac{\partial_0^{\alpha-1} u}{\partial x^{\alpha-1}}(x_1, t_n) \approx \frac{1}{(\Delta x)^{\alpha-1}} \sum_{k=0}^{1} g_k^{\alpha-1} U_{1-k}^n.$$

Therefore, we obtain

$$U_0^{n+1} = U_0^n + \mu_\alpha (g_1^{\alpha-1} U_0^n + g_0^{\alpha-1} U_1^n).$$

A similar problem has been discussed in [60, 139].

Finally, the numerical method is given by

$$U_j^{n+1} = U_j^n + \mu_\alpha \sum_{i=0}^{j+1} g_{j-i+1}^\alpha U_i^n, \quad j = 1, \ldots, N - 1,$$

$$U_0^{n+1} = U_0^n + \mu_\alpha (g_1^{\alpha-1} U_0^n + g_0^{\alpha-1} U_1^n).$$

In this case the matricial form of the problem is

$$\mathbf{U}^{n+1} = (\mathbf{I} + \mu_\alpha \mathbf{A}_{\mathbf{Neu}}) \mathbf{U}^n,$$

with $\mathbf{U}^n = [U_0^n, \ldots, U_{N-1}^n]^T$, $\mathbf{I}$ is the identity matrix and the matrix $\mathbf{A_{Neu}}$ is given by

$$\mathbf{A_{Neu}} = \begin{bmatrix} g_1^{\alpha-1} & g_0^{\alpha-1} & 0 & \cdots & 0 & 0 \\ g_2^{\alpha} & g_1^{\alpha} & g_0^{\alpha} & \cdots & 0 & 0 \\ g_3^{\alpha} & g_2^{\alpha} & g_1^{\alpha} & \cdots & 0 & 0 \\ \vdots & \vdots & \vdots & & \vdots & \vdots \\ g_N^{\alpha} & g_{N-1}^{\alpha} & g_{N-2}^{\alpha} & \cdots & g_2^{\alpha} & g_1^{\alpha} \end{bmatrix}.$$

Highlighted in gray are the entries modified because of the presence of this boundary.

Theorem 6.2 *If $\mu_\alpha \leq 1/\alpha$, then the eigenvalues of the matrix iteration $\mathbf{I} + \mu_\alpha \mathbf{A_{Neu}}$ are less than one.*

Proof See Exercise 6.4.   □

The numerical experiments show that the stability region is given by the necessary stability condition $\mu_\alpha \leq 2^{1-\alpha}$ and not by this more restrictive sufficient stability condition $\mu_\alpha \leq 1/\alpha$ (see Fig. 6.2). The difference between the two curves is very narrow and it is not very significant when the bound of the previous theorem is used in practise to run the numerical experiments inside the stability region of the numerical method.

To develop an higher order numerical method for this type of boundary is a challenge since the overall accuracy may be affected by the approximation we consider at the boundary discrete point. At the moment further work is being developed regarding numerical methods for the fractional diffusion problem with the fractional Neumann boundary. However, many works are still disregarding the fact that in a bounded domain the accuracy of the approximation of the Riemann-Liouville fractional derivative is not the same as when we consider an unbounded domain [162].

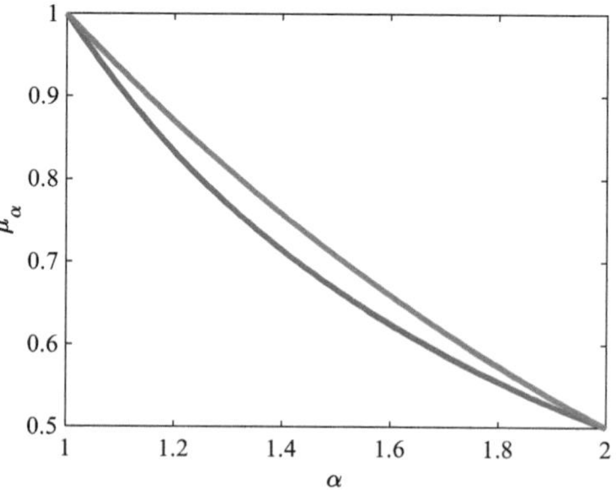

Fig. 6.2 Plots of the regions $\mu_\alpha \leq 1/\alpha$ (smaller region defined by the blue line) and $\mu_\alpha \leq 2^{1-\alpha}$ (larger region defined by the red line)

6.4   A Second Order Scheme for the Symmetric Boundary Wall Problem

In this section, we consider the more general problem that involves the left and right Riemann-Liouville fractional derivatives. The numerical method under study is based on the Riemann-Liouville integral approximation [42, 86, 140] discussed in the fourth chapter. This study has been presented in [70].

Consider the problem defined in the whole real line and denote the integrals in definitions (6.1) and (6.2) respectively by

$$I_+^{2-\alpha} u(x,t) = \frac{1}{\Gamma(2-\alpha)} \int_{-\infty}^{x} u(\xi,t)(x-\xi)^{1-\alpha} d\xi, \tag{6.20}$$

$$I_-^{2-\alpha} u(x,t) = \frac{1}{\Gamma(2-\alpha)} \int_{x}^{\infty} u(\xi,t)(\xi-x)^{1-\alpha} d\xi. \tag{6.21}$$

The domain discretisation is given by $x_k = x_{k-1} + \Delta x$, $k \in \mathbb{Z}$. We approximate the function inside the integrals (6.20) and (6.21) by a linear spline [42, 86, 140], as discussed in the fourth chapter. Hence, we arrive at the following approximations, for the left and right fractional integrals, respectively given by

$$\tilde{I}_+^{2-\alpha} u(x_j,t) = \frac{(\Delta x)^{2-\alpha}}{\Gamma(4-\alpha)} \sum_{m=0}^{\infty} a_m^{\alpha} u(x_{j-m},t), \tag{6.22}$$

$$\tilde{I}_-^{2-\alpha} u(x_j,t) = \frac{(\Delta x)^{2-\alpha}}{\Gamma(4-\alpha)} \sum_{m=0}^{\infty} a_m^{\alpha} u(x_{j+m},t), \tag{6.23}$$

with

$$a_0^{\alpha} = 1, \qquad a_m^{\alpha} = (m+1)^{3-\alpha} - 2m^{3-\alpha} + (m-1)^{3-\alpha}.$$

When we have a reflecting boundary condition at $x = 0$, since the left Riemann-Liouville fractional derivative is modified to (6.7), the modified left fractional integral is now defined by

$$I_{ref}^{2-\alpha} u(x,t) = \frac{1}{\Gamma(2-\alpha)} \int_{0}^{\infty} u(\xi,t)(x+\xi)^{1-\alpha} d\xi$$

$$+ \frac{1}{\Gamma(2-\alpha)} \int_{0}^{x} u(\xi,t)(x-\xi)^{1-\alpha} d\xi. \tag{6.24}$$

The right fractional integral is still defined by (6.21) and therefore can be approximated by (6.23). Following a similar approach as in the open domain, where u

inside the integral is approximated by a linear spline, we obtain the following approximation for the left fractional integral (6.24),

$$\tilde{I}_{ref}^{2-\alpha} u(x_j, t) = \frac{(\Delta x)^{2-\alpha}}{\Gamma(4-\alpha)} \sum_{m=0}^{j} a_m^\alpha u(x_{j-m}, t) + \frac{(\Delta x)^{2-\alpha}}{\Gamma(4-\alpha)} \sum_{m=j+1}^{\infty} a_m^\alpha u(x_{m-j}, t).$$

(6.25)

To discretize the modified left Riemann-Liouville fractional derivative and the right Riemann-Liouville fractional derivative respectively given by

$$\frac{\partial^2}{\partial x^2} I_{ref}^{2-\alpha} u(x, t), \qquad \frac{\partial^2}{\partial x^2} I_-^{2-\alpha} u(x, t),$$

we approximate the second order derivative by a second order central approximation and take in consideration (6.23) and (6.25). For the right Riemann-Liouville fractional derivative, by taking in consideration (6.23), we obtain, as explained in detail in the fourth chapter, the approximation

$$\frac{\partial^2}{\partial x^2} I_-^{2-\alpha} u(x_j, t) \approx \frac{1}{(\Delta x)^\alpha \Gamma(4-\alpha)} \sum_{m=-1}^{\infty} q_m^\alpha u(x_{j+m}, t),$$

with

$$q_{-1}^\alpha = a_0^\alpha, \quad q_0^\alpha = -2a_0^\alpha + a_1^\alpha, \quad q_m^\alpha = a_{m+1}^\alpha - 2a_m^\alpha + a_{m-1}^\alpha, \quad m \geq 1. \quad (6.26)$$

Define

$$\delta_{r,2}^\alpha u(x_j, t) := \frac{1}{\Gamma(4-\alpha)} \sum_{m=-1}^{\infty} q_m^\alpha u(x_{j+m}, t). \tag{6.27}$$

Proceeding in a similar manner, for the modified left Riemann-Liouville fractional derivative, it follows

$$\frac{\partial^2}{\partial x^2} I_{ref}^{2-\alpha} u(x_j, t) \approx \frac{1}{(\Delta x)^\alpha \Gamma(4-\alpha)} \sum_{m=-1}^{j} q_m^\alpha u(x_{j-m}, t)$$

$$+ \frac{1}{(\Delta x)^\alpha \Gamma(4-\alpha)} \sum_{m=j+1}^{\infty} q_m^\alpha u(x_{m-j}, t). \tag{6.28}$$

Define

$$\delta_{l,2,ref}^\alpha u(x_j, t) := \frac{1}{\Gamma(4-\alpha)} \sum_{m=-1}^{j} q_m^\alpha u(x_{j-m}, t) + \frac{1}{\Gamma(4-\alpha)} \sum_{m=j+1}^{\infty} q_m^\alpha u(x_{m-j}, t).$$

(6.29)

We assume a uniform mesh in time and space with

$$t_{n+1} = t_n + \Delta t, \ n \geq 0, \ x_j = x_{j-1} + \Delta x, \ j \in \mathbb{N}.$$

Let U_j^n be the approximated solution of $u(x_j, t_n)$. Consider the Crank-Nicolson scheme to approximate Eq. (6.3) given by

$$\left(1 - \frac{1}{2}\mu_\alpha \delta_{\beta,2,ref}^\alpha\right) U_j^{n+1} = \left(1 + \frac{1}{2}\mu_\alpha \delta_{\beta,2,ref}^\alpha\right) U_j^n + f_j^{n+1/2}, \qquad (6.30)$$

where $f_j^{n+1/2} = (f_j^{n+1} + f_j^n)/2$ and

$$\delta_{\beta,2,ref}^\alpha u(x_j, t) = \frac{1+\beta}{2}\delta_{l,2,ref}^\alpha u(x_j, t) + \frac{1-\beta}{2}\delta_{r,2}^\alpha u(x_j, t). \qquad (6.31)$$

The next result is what we have seen in an open domain. The dependency of the solution u on t is omitted in the following results for the sake of clarity and simplicity.

Theorem 6.3 *Let $u \in C^{(4)}(\mathbb{R})$, $u \in L_\infty(\mathbb{R})$ and u vanishes sufficiently rapidly at infinity together with its derivatives. Then*

$$\frac{\partial^\alpha u}{\partial(-x)^\alpha}(x_j) - \frac{\delta_{r,2}^\alpha u}{\Delta x^\alpha}(x_j) = \epsilon_r(x_j),$$

with $\epsilon_r(x_j) \leq C(\Delta x)^2$, where C does not depend on Δx.

Proof See the proof of Theorem 4.13 in Chap. 4.   □

The next result determines the truncation error for the approximation (6.28) of the modified left Riemann-Liouville fractional derivative.

Theorem 6.4 *Let $u \in C^{(4)}(\mathbb{R})$, $u \in L_\infty(\mathbb{R})$ and such that u vanishes sufficiently rapidly at infinity together with its derivatives. Additionally, u verifies (6.9). Then*

$$\frac{\partial_{ref}^\alpha u}{\partial x^\alpha}(x_j) - \frac{\delta_{l,2,ref}^\alpha u}{(\Delta x)^\alpha}(x_j) = \epsilon_{l,2,ref}(x_j),$$

with $\epsilon_{l,2,ref}(x_j) \leq C_{l,2,ref}(\Delta x)^2$, where $C_{l,2,ref}$ does not depend on Δx.

Proof We have that

$$\tilde{I}_{ref}^{2-\alpha} u(x_j) = \frac{(\Delta x)^{2-\alpha}}{\Gamma(4-\alpha)}\sum_{m=0}^j a_m^\alpha u(x_{j-m}) + \frac{(\Delta x)^{2-\alpha}}{\Gamma(4-\alpha)}\sum_{m=j+1}^\infty a_m^\alpha u(x_{m-j}).$$

Taking in consideration (6.9)

$$\tilde{I}_{ref}^{2-\alpha} u(x_j) = \frac{(\Delta x)^{2-\alpha}}{\Gamma(4-\alpha)} \sum_{m=0}^{j} a_m^{\alpha} u(x_{j-m}) + \frac{(\Delta x)^{2-\alpha}}{\Gamma(4-\alpha)} \sum_{m=j+1}^{\infty} a_m^{\alpha} u(x_{j-m})$$

$$= \frac{1}{\Gamma(4-\alpha)} \sum_{k=-\infty}^{j} \int_{x_{k-1}}^{x_k} s_k(\xi)(x_j - \xi)^{1-\alpha} d\xi,$$

where

$$s_k(\xi) = \frac{x_k - \xi}{\Delta x} u(x_{k-1}) + \frac{\xi - x_{k-1}}{\Delta x} u(x_k).$$

Additionally, by doing a change of variables and taking in consideration the reflecting condition (6.9), we have for the exact value of the integral the following equalities

$$I_{ref}^{2-\alpha} u(x_j) = \int_0^{x_j} u(\xi)(x_j - \xi)^{1-\alpha} d\xi + \int_0^{\infty} u(\xi)(x_j + \xi)^{1-\alpha} d\xi$$

$$= \int_0^{x_j} u(\xi)(x_j - \xi)^{1-\alpha} d\xi - \int_0^{-\infty} u(-\xi)(x_j - \xi)^{1-\alpha} d\xi$$

$$= \frac{1}{\Gamma(4-\alpha)} \sum_{k=-\infty}^{j} \int_{x_{k-1}}^{x_k} u(\xi)(x_j - \xi)^{1-\alpha} d\xi.$$

Therefore

$$I_{ref}^{2-\alpha} u(x_j) - \tilde{I}_{ref}^{2-\alpha} u(x_j) = \frac{1}{\Gamma(4-\alpha)} \sum_{k=-\infty}^{j} \int_{x_{k-1}}^{x_k} (u(\xi) - s_k(\xi))(x_j - \xi)^{1-\alpha} d\xi.$$

From this point on, the proof can follow the same steps as the proof of Theorem 13, presented in the fourth chapter. $\qquad\square$

The next result concerns the stability of the numerical method. Since the problem can be seen as defined in the whole real line, see (6.6), we can use the von Neumann stability analysis.

Theorem 6.5 *The numerical method (6.30) is unconditionally stable.*

Proof The difference operator defined in (6.31) can be rewritten as

$$
\delta_{\beta,2,ref}^{\alpha} U_j^n = \frac{1+\beta}{2}\left[\frac{1}{\Gamma(4-\alpha)}\sum_{m=-1}^{j} q_m^{\alpha} U_{j-m}^n + \frac{1}{\Gamma(4-\alpha)}\sum_{m=j+1}^{\infty} q_m^{\alpha} U_{m-j}^n\right]
$$

$$
+\frac{1-\beta}{2}\frac{1}{\Gamma(4-\alpha)}\sum_{m=-1}^{\infty} q_m^{\alpha} U_{j+m}^n
$$

$$
= \frac{1+\beta}{2}\left[\frac{1}{\Gamma(4-\alpha)}\sum_{m=-1}^{j} q_m^{\alpha} U_{j-m}^n + \frac{1}{\Gamma(4-\alpha)}\sum_{m=j+1}^{\infty} q_m^{\alpha} U_{j-m}^n\right]
$$

$$
+\frac{1-\beta}{2}\frac{1}{\Gamma(4-\alpha)}\sum_{m=-1}^{\infty} q_m^{\alpha} U_{j+m}^n.
$$

The proof of the stability using Fourier analysis consists on inserting a single mode $\kappa^n e^{ij\phi}$ into the numerical scheme (6.30), neglecting the source term, and to verify if the amplification factor κ is not larger than 1, for all $\phi \in [0,\pi]$.

   Inserting a single mode $\kappa^n e^{ij\phi}$ into the numerical scheme (6.30), neglecting the source term, and taking in consideration the previous equality for the reflecting operator, then

$$
\kappa^{n+1} e^{ij\phi} - \frac{1}{2}\mu_\alpha \kappa^{n+1}\left[\frac{1+\beta}{2}\frac{1}{\Gamma(4-\alpha)}\sum_{m=-1}^{\infty} q_m^{\alpha} e^{i(j-m)\phi}\right.
$$

$$
\left.+\frac{1-\beta}{2}\frac{1}{\Gamma(4-\alpha)}\sum_{m=-1}^{\infty} q_m^{\alpha} e^{i(j+m)\phi}\right]
$$

$$
= \kappa^n e^{ij\phi} + \frac{1}{2}\mu_\alpha \kappa^n\left[\frac{1+\beta}{2}\frac{1}{\Gamma(4-\alpha)}\sum_{m=-1}^{\infty} q_m^{\alpha} e^{i(j-m)\phi}\right.
$$

$$
\left.+\frac{1-\beta}{2}\frac{1}{\Gamma(4-\alpha)}\sum_{m=-1}^{\infty} q_m^{\alpha} e^{i(j+m)\phi}\right].
$$

Simplifying $\kappa^n e^{ij\phi}$ on both sides we obtain

$$
\kappa\left\{1 - \frac{1}{2}\mu_\alpha\left[\frac{1+\beta}{2}\frac{1}{\Gamma(4-\alpha)}\sum_{m=-1}^{\infty} q_m^{\alpha} e^{-im\phi}\right.\right.
$$

$$
\left.\left.+\frac{1-\beta}{2}\frac{1}{\Gamma(4-\alpha)}\sum_{m=-1}^{\infty} q_m^{\alpha} e^{im\phi}\right]\right\}
$$

$$= 1 + \frac{1}{2}\mu_\alpha \left[\frac{1+\beta}{2} \frac{1}{\Gamma(4-\alpha)} \sum_{m=-1}^{\infty} q_m^\alpha e^{-im\phi} + \frac{1-\beta}{2} \frac{1}{\Gamma(4-\alpha)} \sum_{m=-1}^{\infty} q_m^\alpha e^{im\phi} \right].$$

If the real part of

$$\left[\frac{1+\beta}{2} \sum_{m=-1}^{\infty} q_m^\alpha e^{-im\phi} + \frac{1-\beta}{2} \sum_{m=-1}^{\infty} q_m^\alpha e^{im\phi} \right]$$

is negative or zero then $|\kappa(\phi)| \le 1$. The real part is given by

$$\left[\frac{1+\beta}{2} \sum_{m=-1}^{\infty} q_m^\alpha \cos(m\phi) + \frac{1-\beta}{2} \sum_{m=-1}^{\infty} q_m^\alpha \cos(m\phi) \right].$$

By the properties of the coefficients q_m^α given in the fourth chapter, we can conclude that the real part is non positive. $\qquad\square$

6.5 The Influence of the Boundaries

We illustrate the effect of the boundaries for different values of α. We display the solution of the fractional diffusion equation, that only involves the left Riemann-Liouville fractional derivative, without the source term and with an initial condition that is an approximation of the Dirac delta function defined as

$$u_0(x) = \frac{1}{\epsilon\sqrt{\pi}} e^{-(x-x_0)^2/\epsilon^2},$$

for a small $\epsilon > 0$. For all figures we have taken $D = 1, \epsilon = 0.1, x_0 = 1$.

In the next figures, we consider two values of α, that is, $\alpha = 1.3$ in Fig. 6.3 and $\alpha = 1.8$ in Fig. 6.4. As we evolve in time, the effect of the boundaries on the solution is quite relevant.

Near the boundary, the behaviour of the solution with the Neumann boundary can be unexpected at first. However, the steady state general solution of the fractional diffusion equation (6.3) is the combination of the functions $x^{\alpha-1}$ and $x^{\alpha-2}$. The solution that goes to zero as x goes to infinity is $x^{\alpha-2}$. This is also the function that better describes the behaviour we observe in the previous figures, near the boundary, since the solution $x^{\alpha-2}$ goes to infinity as x goes to zero.

Near the boundary the solutions behave very differently, highlighting that they represent completely different physical phenomena. Far away from the boundary the behaviour is similar in all three cases: open domain, symmetric reflecting boundary and the fractional Neumann boundary.

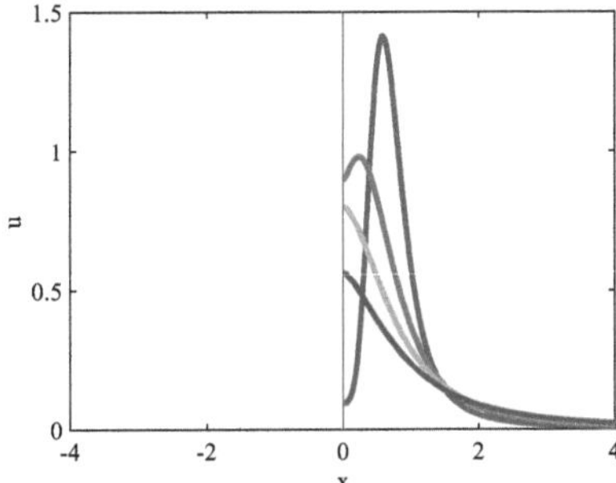 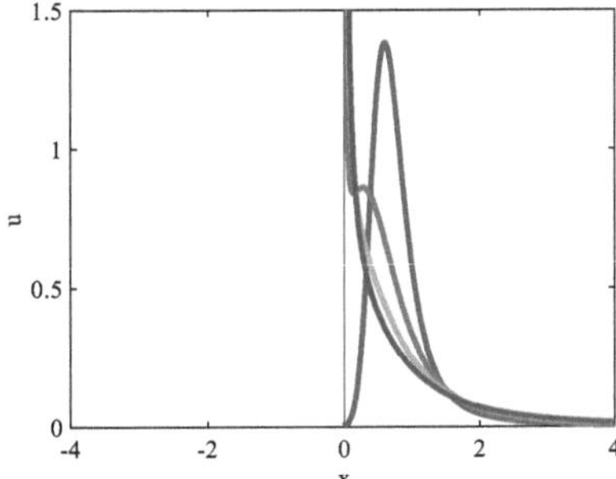

Fig. 6.3 Plots of $u(x, t)$ for $x_0 = 1$, $D = 1$, $\alpha = 1.3$. Symmetric boundary on the left; Neumann boundary on the right. Evolution in time with $t = 0.25, 0.5, 0.75, 1$

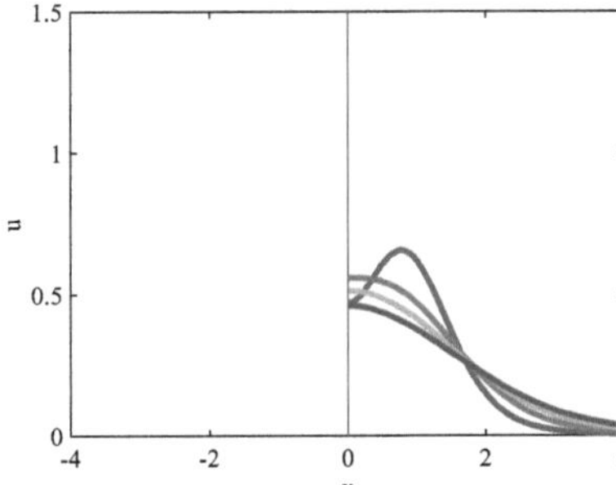 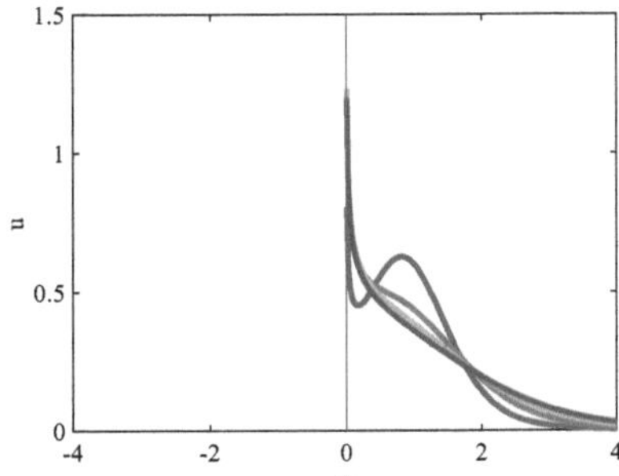

Fig. 6.4 Plots of $u(x, t)$ for $x_0 = 1$, $D = 1$, $\alpha = 1.8$. Symmetric boundary on the left; Neumann boundary on the right. Evolution in time with $t = 0.25, 0.5, 0.75, 1$

6.6   Exercises and Solutions

Exercises

6.1 Consider the heat equation with a left Neumann boundary condition. Prove that the numerical method (6.18), for $\alpha = 2$, is a numerical method suitable for this problem.

6.2 Consider the heat equation with a left Neumann boundary condition. Prove that the explicit Euler method is convergent.

6.3 Prove that

$$\sum_{j=0}^{n} g_j^{\alpha} = g_n^{\alpha-1}.$$

6.4 Prove Theorem 6.2.

6.5 For $n - 1 < \alpha < n$, define the right Riemann-Liouville fractional derivative for $x \leq 1$,

$$\frac{\partial_1^\alpha u}{\partial x^\alpha}(x, t) = \frac{(-1)^n}{\Gamma(n - \alpha)} \frac{\partial^n}{\partial x^n} \int_x^1 u(\xi, t)(\xi - x)^{n-1-\alpha} d\xi.$$

Suppose that we have a problem with a fractional Neumann boundary condition on the right hand side of the domain. We have the following problem, for $\beta \in (-1, 1)$ and $\alpha \in (1, 2)$,

$$\frac{\partial u(x, t)}{\partial t} = D \left(\frac{1 + \beta}{2} \frac{\partial^\alpha u}{\partial x^\alpha}(x, t) + \frac{1 - \beta}{2} \frac{\partial^\alpha u}{\partial(-x)^\alpha}(x, t) \right), \quad x < 1,$$

with

$$\frac{\partial_1^{\alpha-1} u}{\partial x^{\alpha-1}}(1, t) = 0, \quad \text{for all } t > 0.$$

(a) Derive a numerical method to solve this problem using the Grünwald-Letnikov approximation.
(b) Give the iterative matrix of the numerical method by presenting its matricial form.
(c) Discuss the eigenvalues of the iterative matrix.

6.6 Consider the definition, for $n - 1 < \alpha < n$, of the right Riemann-Liouville fractional derivative for $x \leq 1$, given in the previous exercise. Suppose that we have a problem with a fractional Neumann boundary condition on the left and right hand sides of the domain. We have the following problem, for $\beta \in (-1, 1)$ and $\alpha \in (1, 2]$,

$$\frac{\partial u(x, t)}{\partial t} = D \left(\frac{1 + \beta}{2} \frac{\partial^\alpha u}{\partial x^\alpha}(x, t) + \frac{1 - \beta}{2} \frac{\partial^\alpha u}{\partial(-x)^\alpha}(x, t) \right), \quad x \in (0, 1),$$

with

$$\frac{\partial_1^{\alpha-1} u}{\partial x^{\alpha-1}}(0, t) = 0, \quad \text{for all } t > 0,$$

and

$$\frac{\partial_1^{\alpha-1} u}{\partial x^{\alpha-1}}(1, t) = 0, \quad \text{for all } t > 0.$$

(a) Derive a numerical method to solve this problem using the Grünwald-Letnikov approximation.

(b) Give the iterative matrix of the numerical method by presenting its
matricial form.

(c) Discuss the eigenvalues of the iterative matrix.

6.7 Repeat Exercise 6.5 deriving numerical methods based in the spline approximation for the fractional derivatives.

6.8 Repeat Exercise 6.6 deriving numerical methods based in the spline approximation for the fractional derivatives.

Solutions

6.1 We obtain the following discretisation

$$U_j^{n+1} = U_j^n + \mu(U_{j+1}^n - 2U_j^n + U_{j-1}^n), \quad \mu = D\frac{\Delta t}{(\Delta x)^2}.$$

Now for $j = 0$ it reads

$$U_0^{n+1} = U_0^n + \mu(U_1^n - 2U_0^n + U_{-1}^n).$$

Approximating the boundary we obtain

$$U_{-1}^n = U_0^n.$$

Putting it into the difference equation obtained for $j = 0$ we have

$$U_0^{n+1} = U_0^n + \mu(-U_0^n + U_1^n).$$

6.2 We need to use the energy method, since we are not in the real line. However, in this case we have a consistent method. Therefore, it is enough to prove stability to guaranty convergence.

6.3 We can prove the equality by induction. For $n = 1$, taking in consideration that

$$\binom{\alpha}{\beta} = \frac{\Gamma(\alpha + 1)}{\Gamma(\beta + 1)\Gamma(\alpha - \beta + 1)}$$

it is easy to see that

$$\sum_{j=0}^{n} g_j^\alpha = g_0^\alpha + g_1^\alpha = 1 - \alpha = g_1^{\alpha-1}.$$

By the hypothesis of induction, we assume the equality is true for $n - 1$. Therefore, we can write that

$$\sum_{j=0}^{n} g_j^\alpha = \sum_{j=0}^{n-1} g_j^\alpha + g_n^\alpha = g_{n-1}^\alpha + g_n^\alpha.$$

It is now easy to check that

$$g_{n-1}^\alpha + g_n^\alpha = g_{n-1}^\alpha.$$

6.4 Check that the matrix is diagonally dominant and all the elements in the diagonal are negative. Therefore, all the eigenvalues have real part negative. Then, proceed as in Proposition 5.1 of Section 5.4 in the previous chapter. After that, we just need to apply the Gershgorin circle property to the first line of the iterative matrix to conclude that the eigenvalues are less than one under the condition $\alpha \mu_\alpha \leq 1$. We give some details regarding the first line.
By the Gershgorin circle, for the first line, we have

$$|\lambda - (1 + \mu_\alpha g_1^{\alpha-1})| \leq \mu_\alpha.$$

Let $\lambda = a + ib$. Then the previous inequality reads

$$(a - 1 - \mu_\alpha g_1^{\alpha-1})^2 + b^2 \leq \mu_\alpha^2.$$

We have

$$a^2 + b^2 \leq \mu_\alpha^2 + 2a(1 + \mu_\alpha(1 - \alpha)) - (1 + \mu_\alpha(1 - \alpha))^2.$$

After some simplifications we obtain

$$a^2 + b^2 \leq \mu_\alpha^2 \alpha(2 - \alpha) + 2a + 2a\mu(1 - \alpha) - 1 - 2\mu_\alpha + 2\mu_\alpha\alpha.$$

For $\mu_\alpha \alpha \leq 1$ and $a \leq 0$ we have

$$a^2 + b^2 \leq \mu_\alpha(2 - \alpha) + 2a + 2a\mu(1 - \alpha) - 1 - 2\mu_\alpha + 2\mu_\alpha\alpha$$

$$\leq 2a\left[\mu(1 - \alpha) + 1\right] - 1 + \mu_\alpha\alpha$$

$$\leq \mu_\alpha\alpha$$

$$\leq 1.$$

6.5 Present the Crank-Nicolson method or the Euler explicit method and use the results of Chapter 3 to determine the properties of the eigenvalues of the iterative matrix.
6.6 Same approach as Exercise 6.5.
6.7 Same approach as Exercise 6.5.
6.8 Same approach as Exercise 6.5.

The Road Ahead

We conclude this book with a synthesis of topics related to its subject matter and that continue to be the focus of ongoing research. Some of these issues have been highlighted in earlier chapters.

1. Lévy flights and bounded domains from the physical point of view.

Fractional diffusion is linked to Lévy flights, but boundary interactions in this context remain poorly understood. An interesting problem is to rigorously connect stochastic processes with boundary behaviour and to understand the physical implications of different boundary conditions in a probabilistic setting. In addition to the common scenarios of particles being absorbed or reflected, more complex situations may also occur.

2. Well-posedness and regularity of fractional diffusion problems.

The regularity of solutions to fractional diffusion problems exhibits significant differences when compared to the regularity of solutions to classical diffusion problems. A recurring research issue concerns identifying the weakest possible assumptions on the domain and data under which weak and strong solutions exist. Another research question is to understand the behaviour of boundary layers in fractional diffusion problems, particularly in domains with corners or with geometric irregularities.

3. Numerical methods for multidimensional problems.

Fractional diffusion equations involve non-local operators, which make them computationally expensive. This topic was not addressed in the book, as we focused on one-dimensional problems. However, this is of great interest in higher dimensions, since as the number of dimensions increases, so does the computational complexity. Multidimensional problems, tipically in 2D and 3D, are particularly challenging due to memory requirements. This gives rise to a new set of open questions concerning algorithmic strategies. Some of the techniques being devel-

© The Author(s), under exclusive license to Springer Nature Switzerland AG 2026

E. Sousa, *Finite Difference Methods for Fractional Diffusion Equations*, Lecture Notes in Mathematics 2389, https://doi.org/10.1007/978-3-032-11222-4

oped include sparse grid methods, tensor decomposition techniques, reduced-order modeling, Fast Fourier transforms and multigrid methods.

4. Numerical methods for problems with boundary conditions.

There are still many discussions on the proper formulations of boundary conditions for problems involving fractional operators, especially in irregular domains, and on how these conditions affect the uniqueness and stability of solutions. Developing numerical methods for such equations remains computationally challenging, particularly in the treatment of boundary conditions. This challenge is addressed in Chaps. 5 and 6, with Chap. 6 providing a particularly clear illustration of the associated difficulties. Consequently, designing efficient numerical schemes that can accurately capture both the nonlocal nature of the operators and the complexity of the boundary conditions remains an open and active area of research.

5. Adaptive mesh refinement.

Adaptive mesh refinement of the numerical methods can reduce computational cost by focusing resolution where it is needed, such as, near sharp fronts. This is a subject very well documented for classical partial differential equations. However, applying these techniques in the context of nonlocal operators is not straightforward. Determining the best approaches for adaptivity and mesh refinement in fractional diffusion problems, both with and without boundaries, remains an open challenge.

6. High-order schemes.

High-order schemes improve accuracy, but their efficient implementation becomes challenging in general. In the presence of boundaries, difficulties arise because boundary treatments can introduce errors or require additional computational effort. Another interesting question is how to combine these high-order schemes with adaptive mesh refinement. Ensuring stability while doing so remains a significant research challenge.

7. Inverse Problems.

Inverse problems involving fractional diffusion operators can be ill-posed, making them highly sensitive to noise. The most common question is whether coefficients or source terms can be stably recovered from partial boundary measurements. Other key questions concern the controllability and observability of systems governed by fractional diffusion with boundary inputs. Essentially we would like to understand whether we can drive the system from a given initial state to a desired final state by manipulating the data inputs and whether we can determine the full state of the system by measuring outputs only at certain places, such as the boundary.

8. Nonlinear Problems.

Nonlinear fractional partial differential equations, involving both non-linear terms and fractional derivatives, introduce additional complexities. An important question is related to how boundary conditions influence blow-up behavior and long-time asymptotics in nonlinear fractional diffusion problems. From the numerical point of view, it is of interest to explore the use of operator splitting methods and iterative linearization strategies.

References

1. S. Abarbanel, A. Ditkowski, B. Gustafsson, On error bounds of finite difference approximations to partial differential equations – temporal behavior and rate of convergence. J. Sci. Comput. **15**, 79–116 (2000)
2. N.H. Abel, Solution de quelques problèmes à l'aide d'intégrales définies, in *Oeuvres Complètes Christiania*, vol. 1 (Grondahl, Oslo, 1881) , pp. 16–18
3. F. Andreu-Vaillo, J.M. Mazón, J.D. Rossi, J.J. Toledo-Melero, *Nonlocal Diffusion Problems* (American Mathematical Society, Providence, 2010)
4. G.E. Andrews, R. Askey, R. Roy, *Special Functions* (Cambridge University Press, Cambridge, 1999)
5. H.A. Araújo, M.O. Lukin, M.G.E. da Luz, G.M. Viswanathan, F.A.N. Santos, E.P. Raposo, Revisiting Lévy flights on bounded domains: a Fock space approach. J. Stat. Mech. **2020**, P083202 (2020)
6. B. Baeumer, M. Kovács, H. Sankaranarayanan, Fractional partial differential equations with boundary conditions. J. Differ. Equ. **264**, 1377–1410 (2018)
7. B. Baeumer, M. Luks, M.M. Meerschaert, Space-time fractional Dirichlet problems. Math. Nachr. **291**, 2516–2535 (2018)
8. B. Baeumer, M. Kovács, M.M. Meerschaert, H. Sankaranarayanan, Boundary conditions for fractional diffusion. J. Comput. Appl. Math. **336**, 408–424 (2018)
9. Z.Z. Bai, K.Y. Lu, Fast matrix splitting preconditioners for higher dimensional spatial fractional diffusion equations. J. Comput. Phys. **404**, 109117 (2020)
10. O.G. Bakunin, *Turbulence and Diffusion* (Springer, Berlin, 2008)
11. S.G. Bardeji, I.N. Figueiredo, E. Sousa, Optical flow with fractional order regularization: variational model and solution method. Appl. Numer. Math. **114**, 188–200 (2017)
12. B. Barrios, I. Peral, F. Soria, E. Valdinoci, A Widder's type theorem for the heat equation with nonlocal diffusion. Arch. Ration. Mech. Anal. **213**, 629–650 (2014)
13. R.G. Bartle, *The Elements of Integration and Lebesgue Measure* (Wiley, Hoboken, 1995)
14. D.A. Benson, S.W. Wheatcraft, M.M. Meerschaert, Application of a fractional advection-dispersion equation. Water Resour. Res. **36**, 1403–1412 (2000)
15. M. Bernkopf, A history of infinite matrices. Arch. Hist. Exact Sci. **4**, 308–358 (1968)
16. M. Bonforte, Y. Sire, J.L. Vázquez, Optimal existence and uniqueness theory for the fractional heat equation. Nonlinear Anal. **153**, 142–168 (2017)
17. S. Borak, W. Hardle, R. Weron, Stable distributions, in *Statistical Tools for Finance and Insurance* (Springer, Berlin, 2005)
18. A. Böttcher, S.M. Grudsky, *Spectral Properties of Banded Toeplitz Matrices* (SIAM, Philadelphia, 2005)

© The Author(s), under exclusive license to Springer Nature Switzerland AG 2026

E. Sousa, *Finite Difference Methods for Fractional Diffusion Equations*, Lecture Notes in Mathematics 2389, https://doi.org/10.1007/978-3-032-11222-4

19. D. Brockmann, L. Hufnage, T. Geisel, The scalling laws of human travel. Nature **439**, 462–465 (2006)

20. C. Bucur, E. Valdinoci, *Nonlocal Diffusion and Applications* (Springer, Berlin, 2016)

21. S.V. Buldyrev, S. Havlin, A.Y. Kazakov, M.G.E. da Luz, E.P. Raposo, H.E. Stanley, G.M. Viswanathan, Average time spent by Lévy flights and walks on an interval with absorbing boundaries. Phys. Rev. **64**, 041108 (2001)

22. N. Burch, R.B. Lehoucq, Continuous-time random walks on bounded domains. Phys. Rev. E **83**, 012105 (2011)

23. P.L. Butzer, R.J. Nessel, *Fourier Analysis and Approximation* (Birkhauser Verlag, Basel, 1971)

24. L.A. Caffarelli, A. Vasseur, Drift diffusions equations with fractional diffusion and the quasi-geostrophic equation, Ann. Math. **171**, 1903–1930 (2010)

25. M. Caputo, Linear models of dissipation whose Q is almost frequency independent-II. Geophys. J. R. Astron. Soc. **13**, 529–539 (1967)

26. M. Caputo, F. Mainardi, A new dissipation model based on memory mechanism. Fract. Calc. Appl. Anal. **10**, 309–324 (2007); Reprinted from Pure Appl. Geophys. 9, 134–147 (1971)

27. A.R. Carella, C.A. Dorao, Least-squares spectral method for the solution of a fractional advection-dispersion equation. J. Comput. Phys. **232**, 33–45 (2013)

28. P. Castillo, S. Gómez, Conservative local discontinuous Galerkin method for the fractional Klein-Gordon-Schrödinger system with generalized Yukawa interaction. Numer. Algor. **84**, 407–425 (2020)

29. C. Çelik, M. Duman, Crank-Nicolson method for the fractional diffusion equation with Riesz fractional derivative. J. Comput. Phys. **231**, 1743–1750 (2012)

30. R.H. Chan, A. Lanza, S. Morigi, F. Sgallari, An adaptive strategy for the restoration of textured images using fractional order regularization. Numer. Math. Theor. Meth. Appl. **6**, 276–296 (2013)

31. P. Chakraborty, M.M. Meerschaert, C.Y. Lim, Parameter estimation for fractional transport: a particle-tracking approach. Water Resour. Res. **45**, W10415 (2009)

32. F. Chatelin, *Eigenvalue of Matrices* (SIAM, Philadelphia, 2012)

33. A.V. Chechkin, R. Metzler, V.Y. Gonchar, J. Klafter, L.V. Tanatarov, First passage and arrival time densities for Lévy flights and the failure of the method of images. J. Phys. A. Math. Gen. **36**, L537 (2003)

34. W. Chen, A speculative study of 2/3-order fractional Laplacian modeling of turbulence: some thoughts and conjectures. Chaos **16**, 023126 (2006)

35. H. Chen, F. Holland, M. Stynes, An analysis of the Grünwald-Letnikov scheme for initial-value problems with weakly singular solutions. Appl. Numer. Math. **139**, 52–61 (2019)

36. P. Constantin, J. Wu, Behaviour of solutions of 2D quasi-geostrophic equations. SIAM J. Math. Anal. **30**, 937–948 (1999)

37. N. Cusimano, K. Burrage, I. Turner, D. Kay, On reflecting boundary conditions for space-fractional equations on a finite interval: proof of the matrix transfer technique. Appl. Math. Model. **42**, 554–565 (2017)

38. P.J. Davis, *Circulant Matrices* (Chelsea Publishing, New York, 1994)

39. O. Defterli, M. D'Elia, Q. Du, M. Gunzburger, R. Lehoucq, M.M. Meerschaert, Fractional diffusion in bounded domains. Fract. Calc. Appl. Anal. **18**, 342–360 (2015)

40. Z.-Q. Deng, J.L.M.P. de Lima, M.I.P. de Lima, V.P. Singh, A fractional dispersion model for overland solute transport. Water Resour. Res. **42**, W03416 (2006)

41. M. D'Elia, M. Gunzburger, The fractional Laplacian operator on bounded domains as a special case of the nonlocal diffusion operator. Comput. Math. Appl. **66**, 1245–1260 (2013)

42. K. Diethelm, N.J. Ford, A.D. Freed, Detailed error analysis for a fractional Adams method. Numer. Algor. **36**, 31–52 (2004)

43. K. Diethelm, *The Analysis of Fractional Differential Equations*. Lecture Notes in Mathematics (Springer, Berlin, 2010)

44. S. Dipierro, X. Ros-Oton, E. Valdinoci, Nonlocal problems with Neumann boundary conditions. Rev. Mat. Iberoam. **33**, 377–416 (2017)

45. M. Donatelli, R. Krause, M. Mazza, K. Trotti, Multigrid preconditioners for anisotropic space-fractional diffusion equations. Adv. Comput. Math. **46**, 49 (2020)

46. O. Dovgoshey, O. Martio, V. Ryazanov, M. Vuorinen, The Cantor function. Expo. Math. **24**, 1–37 (2006)

47. A.A. Dubkov, A. La Cognata, B. Spagnolo, The problem of analytical calculation of barrier crossing characteristics for Lévy flights. J. Stat. Mech. **2009**, P01002 (2009)

48. B. Dybiec, E. Gudowska-Nowak, P. Hanggi, Lévy-Brownian motion on finite intervals: mean first passage time analysis. Phys. Rev. E **73**, 046104 (2006)

49. B. Dybiec, E. Gudowska-Nowak, E. Barkai, A.A. Dubkov, Lévy flights versus Lévy walks in bounded domains. Phys. Rev. E **95**, 052102 (2017)

50. A. Einstein, Über die von der molekularkinetischen Theorie der Wärme geforderte Bewegung von in ruhenden Flüssigkeiten suspendierten Teilchen, A.Ann. Phys. **17**, 549–560 (1905)

51. D. Elliott, An asymptotic analysis of two algorithms for certain Hadamard finite-part integrals. IMA J. Numer. Anal. **13**, 445–462 (1993)

52. V.J. Ervin, Regularity of the solution to fractional diffusion, advection, reaction equations in weighted Sobolev spaces. J. Differ. Equ. **278**, 294–325 (2021)

53. V.J. Ervin, J.P. Roop, Variational formulation for the stationary fractional advection dispersion equation. Numer. Methods Partial Differ. Equ. **22**, 507–760 (2006)

54. V.J. Ervin, J.P. Roop, Regularity of the solution to 1-D fractional order diffusion equations. Math. Comput. **87**, 2273–2294 (2018)

55. Q. Fang, Convergence of finite difference methods for convection-diffusion problems with singular solutions. J. Comput. Appl. Math. **152**, 119–131 (2003)

56. R. Fernandez, J.L. Mateos, O. Miramontes, G. Cocho, H. Larralde, B. Ayala-Orozco, Lévy walk patterns in the foraging movements of spider monkeys. Behav. Ecol. Soc. **55**, 223–230 (2004)

57. H.C. Fogedby, Lévy Flights in random environments. Phys. Rev. Lett. **73**, 2517 (1994)

58. K. Górska, K.A. Penson, Lévy stable two-sided distributions: exact and explicit densities for asymmetric case. Phys. Rev. E **83**, 061125 (2011)

59. J.L. Gracia, M. Stynes, Formal consistency versus actual convergence rates of difference schemes for fractional derivative boundary value problems. Fract. Calc. Appl. Anal. **18**, 419–436 (2015)

60. J.L. Gracia, M. Stynes, A finite difference method for an initial–boundary value problem with a Riemann-Liouville-Caputo spatial fractional derivative. J. Comput. Appl. Math. **381**, 113020 (2021)

61. J.L. Gracia, E. O'Riordan, M. Stynes, Convergence analysis of a finite difference scheme for a two-point boundary value problem with a Riemann-Liouville-Caputo fractional derivative. Bit Numer. Math. **60**, 411–439 (2020)

62. R.M. Gray, Toeplitz and circulant matrices: a review. Found. Trends Commun. Inf. Theory **2**, 155–239 (2006)

63. S. Guo, L. Mei, Z. Zhang, C. Li, M. Li, Y. Wang, A linearized finite difference/spectral-Galerkin scheme for three-dimensional distributed order time-space fractional nonlinear reaction-diffusion-wave equation: numerical simulations of Gordon-type solitons. Comput. Phys. Commun. **252**, 107144 (2020)

64. B. Gustafsson, The convergence rate for difference approximations to mixed initial boundary value problems. Math. Comput. **29**, 396–406 (1975)

65. B. Gustafsson, The convergence rate for difference approximations to general mixed initial boundary value problems. SIAM J. Numer. Anal. **18**, 179–190 (1981)

66. Z. Hao, H. Zang, Optimal regularity and error estimates of a spectral Galerkin method for fractional advection-diffusion-reaction equations. SIAM J. Numer. Anal. **58**, 211–233 (2020)

67. G. Huang, Q. Huang, H. Zhan, Evidence of one-dimensional scale-dependent fractional advection-dispersion. J. Contam. Hydrol. **85**, 53–71 (2006)

68. Y. Huang, A. Oberman, Numerical methods for the fractional Laplacian: a finite difference-quadrature approach. SIAM J. Numer. Anal. **52**, 3056–3084 (2014)

69. S. Jespersen, R. Metzler, H.C. Fogedby, Lévy flights in external force fields: Langevin and fractional Fokker-Planck equations and their solutions. Phys. Rev. E. **59**, 2736 (1999)

70. C. Jesus, E. Sousa, Numerical solutions for asymmetric Lévy flights. Numer. Algor. **87**, 967–999 (2021)

71. W. Jiang, Y. Lin, Approximate solution of the fractional advection dispersion equation. Comput. Phys. Commun. **181**, 557–561 (2010)

72. B. Jin, R. Lazarov, J. Pasciak, Z. Zhou, Error analysis of a finite element method for the space-fractional parabolic equation. SIAM J. Numer. Anal. **52**, 2272–2294 (2014)

73. B. Jin, R. Lazarov, J. Pasciak, W. Rundell, Variational formulation of problems involving fractional order differential operators. Math. Comput. **84**, 2665–2700 (2015)

74. J.F. Kelly, H. Sankaranarayanan, M.M. Meerschaert, Boundary conditions for two-sided fractional diffusion. J. Comput. Phys. **376**, 1089–1107 (2019)

75. A.A. Kilbas, H.M. Srivastava, J.J. Trujillo, *Theory and Applications of Fractional Differential Equations* (Elsevier, Amsterdam, 2006)

76. A. Kiselev, F. Nazarov, A. Volberg, Global well-posedness for the critical 2D dissipative quasi-geostrophic equation. Invent. Math. **167**, 445–453 (2007)

77. N. Krepysheva, L. Di Pietro, M.C. Néel, Fractional diffusion and reflective boundary condition. Phys. A **368**, 355–361 (2006)

78. N. Krepysheva, L. Di Pietro, M.C. Néel, Space-fractional advection-diffusion and reflective boundary condition. Phys. Rev. E **73**, 021104 (2006)

79. P.D. Lax, R.D. Richtmyer, Survey of the stability of linear finite difference equations. Commun. Pure Appl. Math. **9**, 267–293 (1956)

80. P.D. Lax, B. Wendroff, Difference schemes for hyperbolic equations with high order of accuracy. Commun. Pure Appl. Math. **17**, 381–398 (1964)

81. B.P. Leonard, A stable and accurate convective modelling procedure based on quadratic upstream interpolation. Comput. Methods Appl. Mech. Eng. **19**, 59–98 (1979)

82. C. Li, W. Deng, High order schemes for the tempered fractional diffusion equations. Adv. Comput. Math. **42**, 543–572 (2016)

83. C. Li, C. Tao, On the fractional Adam's method. Comput. Math. Appl. **58**, 1573–1588 (2009)

84. C. Li, F. Zheng, Finite difference methods for fractional diffusion equations. Int. J. Bifurcat. Chaos **22**, 1230014 (2012)

85. C. Li, F. Zeng, *Numerical Methods for Fractional Calculus* (CRC Press, Boca Raton, 2015)

86. C. Li, A. Chen, J. Ye, Numerical approaches to fractional calculus and fractional ordinary differential equation. J. Comput. Phys. **230**, 3352–3368 (2011)

87. H. Liu, X. Zheng, H. Wang, H. Fu, Error estimate of finite element approximation for two-sided space-fractional evolution equation with variable coefficient. J. Sci. Comput. **90**, 15 (2022)

88. A. Lischke, J.F. Kelly, M.M. Meerschaert, Mass-conserving tempered fractional diffusion in a bounded interval. Fract. Calc. Appl. Anal. **22**, 1561–1595 (2019)

89. C. Lubich, Disctretized fractional calculus. SIAM J. Math. Anal. **17**, 704–719 (1986)

90. C. Lubich, Fractional linear multistep methods for Abel-Volterra integral equations of the first kind. IMA J. Numer. Anal. **7**, 97–106 (1987)

91. R.L. Magin, O. Abdullah, D. Baleanu, X.J. Zhou, Anomalous diffusion expressed through fractional order differential operators in the Bloch-Torrey equation. J. Magn. Reson. **190**, 255–270 (2008)

92. F. Mainardi, Y. Lucho, G. Pagnini, The fundamental solution of the space-time fractional diffusion equation. Fract. Calc. Appl. Anal. **4**, 153–192 (2001)

93. F. Mainardi, G. Pagnini, R.K. Saxena, Fox H functions in fractional diffusion. J. Comput. Appl. Math. **178**, 321–331 (2005)

94. B.B. Mandelbrot, *The Fractal Geometry of Nature* (W.H. Freeman and Company, New York, 1982)

95. A.M. Mathai, R.K. Saxena, H.J. Haubold, *The H-Function: Theory and Applications* (Springer, Berlin, 2009)

96. N. Matsunaga, T. Yamamoto, Superconvergence of the Shortley-Weller approximation for Dirichlet problems. J. Comput. Appl. Math. **116**, 263–273 (2000)
97. M.M. Meerschaert, C. Tadjeran, Finite difference approximations for fractional advection-dispersion flow equations. J. Comput. Appl. Math. **172**, 65–77 (2004)
98. R. Metzler, J. Klafter, The random walk's guide to anomalous diffusion: a fractional dynamics approach. Phys. Rep. **339**, 1–77 (2000)
99. R. Metzler, J. Klafter, Boundary value problems for fractional diffusion equations. Phys. A Stat. Mech. Appl. **278**, 107–125 (2000)
100. R. Metzler, J. Klafter, The restaurant at the end of the random walk: recent developments in the description of anomalous transport by fractional dynamics. J. Phys. A Math. Gen. **37**, R161 (2004)
101. R. Metzler, A.V. Chechkin, J. Klafter, Levy statistics and anomalous transport: levy flights and subdiffusion, in: *Encyclopedia of Complexity and Systems Science*, ed. by R. Meyers (Springer, New York, 2009)
102. K.W. Morton, *Numerical Solution of Convection Diffusion Problems* (Chapman and Hall, London, 1996)
103. K.W. Morton, D.F. Mayers, *Numerical Solution of Partial Differential Equations* (Cembridge University Press, Cambridge, 1994)
104. K.B. Oldham, J. Spanier, *The Fractional Calculus* (Dover, Mineola, 1974)
105. A. Padash, A.V. Chechkin, B. Dybiec, I. Pavlyukevich, B. Shokri, R. Metzler, First passage properties of asymmetric Lévy flights. J. Phys. A Math. Theor. **52**, 454004 (2019)
106. H.-K. Pang, H.-W. Sun, Multigrid method for fractional diffusion equations. J. Comput. Phys. **231**, 693–703 (2012)
107. K.A. Penson, K. Górska, Exact and explicit probability densities for one-sided Lévy stable distributions. Phys. Rev. Lett. **105**, 210604 (2010)
108. C. Piret, E. Hanert, A radial basis functions method for fractional diffusion equations. J. Comput. Phys. **238**, 71–81 (2013)
109. I. Podlubny, *Fractional Differential Equations* (Academic Press, Cambridge, 1999)
110. I. Podlubny, R. Magin, I. Trymorush, Niels Henrik Abel and the birth of fractional calculus. Fract. Calc. Appl. Anal. **20**, 1068–1075 (2017)
111. B. Podobnik, A. Valentincic, D. Horvatic, H.E. Stanley, Asymmetric Lévy flight in financial ratios. P. Natl. Acad. Sci. USA **108**, 17883–17888 (2011)
112. R. Renn, Einstein's invention of Brownian motion. Ann. Phys. **14**, 23–37 (2005)
113. R.D. Richtmyer, K.W. Morton, *Difference Methods for Initial Value Problems* (Wiley, Hoboken, 1967)
114. P.D. Romero, F.V. Candela, Blind deconvolution models regularized by fractional powers of the Laplacian. J. Math. Imaging Vis. **32**, 181–191 (2008)
115. X. Ros-Oton, J. Serra, The Dirichlet problem for the fractional Laplacian: regularity up to the boundary. J. Math. Pures Appl. **101**, 275–302 (2014)
116. B. Ross (Ed.), *Fractional Calculus and its Applications*. Lecture Notes in Mathematics, vol. 457 (Springer, Berlin, 1975)
117. B. Ross, The development of fractional calculus 1695–1900. Historia Math. **4**, 75–89 (1977)
118. A.I. Saichev, G.M. Zaslavsky, Fractional kinetic equations: solutions and applications. Chaos **7**, 753–764 (1997)
119. S.G. Samko, A.A. Kilbas, O.I. Marichev, *Fractional Integrals and Derivatives: Theory and Applications* (Gordon and Breach Science Publishers, London, 1993)
120. G. Samorodnitsky, M.S. Taqqu, *Stable Non-Gaussian Random Processes: Stochastic Models with Infinite Variance* (Chapman and Hall/CRC, Boca Raton, 1994)
121. M. Schmiedeberg, V.Y. Zaburdaev, H. Stark, On moments and scaling regimes in anomalous random walks. J. Stat. Mech. **2009**, P12020 (2009)
122. W.R. Schneider, Stable distributions: fox function representation and generalization, in: *Stochastic Processes in Classical and Quantum Systems*, ed. by S. Albeverio, G. Casati, D. Merlini. Lecture Notes in Physics, vol. 262 (Springer, Berlin, 1986)

123. S. Shen, F. Liu, V. Ahn, I. Turner, J. Chen, A characteristic difference method for the variable-order fractional advection-diffusion equation. J. Appl. Math. Comput. **42**, 371–386 (2013)

124. M.F. Shlesinger, B.J. West, J. Klafter, Lévy dynamics of enhanced diffusion: application to turbulence. Phys. Rev. Lett. **58**, 1100–1103 (1987)

125. M.F. Shlesinger, G.M. Zaslavsky, J. Klafter, Strange kinetics. Nature **363**, 31–37 (1993)

126. G.D. Smith, *Numerical Solution of Partial Differential Equations: Finite Difference Methods* (Oxford University Press, Oxford, 1985)

127. E. Sousa, The controversial stability analysis. J. Appl. Math. Comput. **145**, 777–794 (2003)

128. E. Sousa, High order schemes and numerical boundary conditions. Comput. Methods Appl. Mech. Eng. **196**, 4444–4457 (2007)

129. E. Sousa, Finite difference approximations for a fractional advection diffusion problem. J. Comput. Phys. **228**, 4038–4054 (2009)

130. E. Sousa, Numerical approximations for fractional diffusion equations via splines. Comput. Math. Appl. **62**, 938–944 (2011)

131. E. Sousa, A second order explicit finite difference method for the fractional advection diffusion equation. Comput. Math. Appl. **64**, 3141–3152 (2012)

132. E. Sousa, How to approximate the fractional derivative of order $1 < \alpha < 2$. Int. J. Bifurcat. Chaos **22**, 1250075 (2012)

133. E. Sousa, Numerical solution of a model for turbulent diffusion. Int. J. Bifurcation Chaos **23**, 1350166 (2013)

134. E. Sousa, A close look at fractional derivatives approximations, in *Fractional Differentiation and Its Applications (ICFDA), 2014 International Conference on IEEE Xplore Abstract* (2014)

135. E. Sousa, An explicit high order method for fractional advection diffusion equations. J. Comput. Phys. **278**, 257–274 (2014)

136. E. Sousa, Consistency analysis of the Grünwald-Letnikov approximation in a bounded domain. IMA J. Numer. Anal. **42**, 2771–2793 (2022)

137. E. Sousa, Convergence of consistent and inconsistent schemes for fractional diffusion problems with boundaries. Adv. Comput. Math. **48**, 68 (2022)

138. E. Sousa, The convergence rate for difference approximations to fractional boundary value problems. J. Comput. Appl. Numer. Math. **415**, 114486 (2022)

139. E. Sousa, Fractional diffusion problems with reflecting boundaries. Lect. Notes Comput. Sci. **13952**, 164–171 (2024)

140. E. Sousa, C. Li, A weighted finite difference method for the fractional diffusion based on the Riemann-Liouville derivative. Appl. Numer. Math. **90**, 22–37 (2015)

141. R. Stern, F. Effenberger, H. Fichtner, T. Schafer, The space fractional diffusion advection equation: analytical solution and critical assessment of numerical solutions. Fract. Calc. Appl. Anal. **17**, 171–190 (2013)

142. P.R. Stinga, J.L. Torrea, Extension problem and Harnack's inequality for some fractional operators. Commun. Partial. Differ. Equ. **35**, 2092–2122 (2010)

143. M. Svärd, J. Nordström, On the convergence rates of energy-stable finite-difference schemes. J. Comput. Phys. **397**, 108819 (2019)

144. B. Szekeres, F. Izsák, A finite difference method for fractional diffusion equations with Neumann boundary conditions. Open Math. **13**, 581–600 (2015)

145. C. Tadjeran, M.M. Meerschaert, H.-P. Scheffler, A second-order accurate numerical approximation for the fractional diffusion equation. J. Comput. Phys. **213**, 205–213 (2006)

146. J.W. Thomas, *Numerical Partial Differential Equtaions* (Springer, New York, 1995)

147. B.H. Thorsen, The eigenvalues of an infinite matrix. College Math. J. **31**, 107–110 (2000)

148. V.K. Tuan, R. Gorenflo, Extrapolation to the limit for numerical fractional differentiation. Z. Agnew. Math. Mech. **75**, 646–648 (1995)

149. V.V. Uchaikin, V.M. Zolotarev, *Chance and Stability. Stable Distributions and their Applications* (De Gruyter, Berlin, 1999)

150. B.P. van Milligen, I. Calvo, R. Sanchez, Continuous random walks in finite domains and general boundary conditions: some formal considerations. J. Phys. A Math. Theor. **41**, 215004 (2008)
151. R.S. Varga, *Matrix Iterative Analysis* (Springer, Kent, 2009)
152. J.L. Vázquez, A. de Pablo, F. Quirós, A. Rodríquez, Classical solutions and higher regularity for nonlinear fractional diffusion equations. J. Eur. Math. Soc. **19**, 1949–1975 (2017)
153. S. Vong, P. Lyu, X. Chen, S.L. Lei, High order finite difference method for time-space fractional differential equations with Caputo and Riemann-Liouville derivatives. Numer. Algor. **72**, 195–210 (2016)
154. H. Wang, D. Yang, Wellposedness of variable-coefficient conservative fractional elliptic differential equations. SIAM J. Numer. Anal. **51**, 1088–1107 (2013)
155. H. Wang, D. Yang, Wellposedness of Neumann boundary-value problems of space-fractional differential equations. Fract. Calc. Appl. Anal. **20**, 1356–1381 (2017)
156. A. Wardak, First passage leapovers of Lévy flights and the proper formulation of absorbing boundary conditions. J. Phys. A Math. Theor. **53**, 375001 (2020)
157. F.F. Warming, B.J. Hyett, The modified equation approach to the stability and accuracy analysis of finite difference methods. J. Comput. Phys. **14**, 159–179 (1974)
158. N.P. Waterson, H. Deconinck, Design principles for bounded higher-order convection schemes-a unified approach. J. Comput. Phys. **224**, 182–207 (2007)
159. B.J. West, Fractional calculus view of complexity: a tutorial. Rev. Mov. Phys. **86**, 1169 (2014)
160. B.J. West, P. Grigolini, R. Metzler, T.F. Nonnenmacher, Fractional diffusion and Lévy stable processes. Phys. Rev. E **55**, 99–106 (1997)
161. Y. Wu, R.A. Falconer, A mass conservative 3D numerical model for predicting solute fluxes in estuarine waters.Adv. Water Resour. **23**, 531–543 (2000)
162. C. Xie, S. Fang, Efficient numerical methods for Riesz space-fractional diffusion equations with fractional Neumann boundary conditions. Appl. Numer. Math. **176**, 1–18 (2022)
163. T. Yamamoto, Convergence of consistent and inconsistent finite difference schemes and acceleration technique. J. Comput. Appl. Math. **140**, 849–866 (2002)
164. B. Yin, Y. Liu, H. Li, Necessity of introducing non-integer shifted parameters by constructing high accuracy finite difference algorithms for a two-sided space-fractional advection-diffusion model. Appl. Math. Lett. **105**, 106347 (2020)
165. M.A. Zaky, A.S. Hendy, J.E. Macías-Díaz, Semi-implicit Galerkin-Legendre spectral schemes for nonlinear time-space fractional diffusion-reaction equations with smooth and nonsmooth solutions. J. Sci. Comput. **82**, 13 (2020)
166. G.M. Zaslavsky, Chaos, fractional kinetics, and anomalous transport. Phys. Rep. **371**, 461–580 (2002)
167. F. Zeng, C. Li, F. Liu, I. Turner, The use of finite difference/element approaches for solving the time-fractional subdiffusion equation. SIAM J. Sci. Comput. **35**, A2976–A3000 (2013)
168. F. Zeng, Z. Zhang, G.E. Karniadakis, Second-order numerical methods for multi-term fractional differential equations: smooth and non-smooth solutions. Comput. Methods Appl. Mech. Eng. **327**, 478–502 (2017)
169. X. Zhang, L. Mauchao, J.W. Crawford, I.M. Young, The impact of boundary on the fractional advection-dispersion equation for solute transport in soil: defining the fractional dispersive flux with the Caputo derivatives. Adv. Water Resour. **30**, 1205–1217 (2007)
170. J. Zhang, Z. Wei, L. Xiao, Adaptive fractional-order multi-scale method for image denoising. J. Math. Imag. Vis. **43**, 39–49 (2012)
171. Y. Zhang, X. Yu, X. Li, Kelly, J.F., H. Sun, C. Zhen, Impact of absorbing and reflective boundaries on fractional derivative models: quantification, evaluation and application. Adv. Water Resour. **128**, 129–144 (2019)
172. H. Zhang, F. Liu, S. Chen, M. Shen, A fast and high accuracy numerical simulation for a fractional Black-Scholes model on two assets. Ann. Appl. Math. **36**, 91–110 (2020)
173. Q. Zhang, J.S. Hesthaven, Z.-Z. Sun, Y. Ren, Pointwise error estimate in difference setting for the two-dimensional nonlinear fractional complex Ginzburg-Landau equation. Adv. Comput. Math. **47**, 35 (2021)

174. Z. Zhao, C. Li, Fractional difference/finite element approximations for the time-space fractional telegraph equation. Appl. Math. Comput. **219**, 2975–2988 (2012)
175. X. Zheng, V.J. Ervin, H. Wang, Optimal Petrov-Galerkin spectral approximation method for the fractional diffusion, advection, reaction equation on a bounded interval. J. Sci. Comput. **86**, 1–22 (2021)
176. L. Zhou, H.M. Selim, Application of the fractional advection-dispersion equation in porous media. Soil Sci. Soc. Am. J. **67**, 1079–1084 (2004)
177. X.J. Zhou, Q. Gao, O. Abdullah, R.L. Magin, Studies of anomalous diffusion in the human brain using fractional order calculus. Magnet. Reson. Med. **63**, 562–569 (2010)
178. A. Zoia, A. Rosso, M. Kardar, Fractional Laplacian in bounded domains. Phys. Rev. E **76**, 021116 (2007)
179. G. Zumofen, J. Klafter, Absorbing boundary in one-dimensional anomalous transport. Phys. Rev. E **51**, 2805 (1995)

Index

forwarded to the LNM Editorial Board, this is very helpful. If no reports are forwarded or if other questions remain unclear in respect of homogeneity etc, the series editors may wish to consult external referees for an overall evaluation of the volume.

5. Manuscripts should in general be submitted in English. Final manuscripts should contain at least 100 pages of mathematical text and should always include

   - a table of contents;
   - an informative introduction, with adequate motivation and perhaps some historical remarks: it should be accessible to a reader not intimately familiar with the topic treated;
   - a subject index: as a rule this is genuinely helpful for the reader.
   - For evaluation purposes, manuscripts should be submitted as pdf files.

6. Careful preparation of the manuscripts will help keep production time short besides ensuring satisfactory appearance of the finished book in print and online. After acceptance of the manuscript authors will be asked to prepare the final LaTeX source files (see LaTeX templates online: https://www.springer.com/gb/authors-editors/book-authors-editors/manuscriptpreparation/5636) plus the corresponding pdf- or zipped ps-file. The LaTeX source files are essential for producing the full-text online version of the book, see http://link.springer.com/bookseries/304 for the existing online volumes of LNM). The technical production of a Lecture Notes volume takes approximately 12 weeks. Additional instructions, if necessary, are available on request from lnm@springer.com.

7. Authors receive a total of 30 free copies of their volume and free access to their book on SpringerLink, but no royalties. They are entitled to a discount of 33.3 % on the price of Springer books purchased for their personal use, if ordering directly from Springer.

8. Commitment to publish is made by a *Publishing Agreement*; contributing authors of multiauthor books are requested to sign a *Consent to Publish form*. Springer-Verlag registers the copyright for each volume. Authors are free to reuse material contained in their LNM volumes in later publications: a brief written (or e-mail) request for formal permission is sufficient.

Addresses:
Professor Jean-Michel Morel, CMLA, École Normale Supérieure de Cachan, France
E-mail: moreljeanmichel@gmail.com

Professor Bernard Teissier, Equipe Géométrie et Dynamique,
Institut de Mathématiques de Jussieu – Paris Rive Gauche, Paris, France
E-mail: bernard.teissier@imj-prg.fr

Springer: Ute McCrory, Mathematics, Heidelberg, Germany,
E-mail: lnm@springer.com

MIX
Papier aus verantwortungsvollen Quellen
Paper from responsible sources
FSC® C105338

If you have any concerns about our products,
you can contact us on
ProductSafety@springernature.com

In case Publisher is established outside the EU,
the EU authorized representative is:
**Springer Nature Customer Service Center GmbH
Europaplatz 3, 69115 Heidelberg, Germany**

Printed by Libri Plureos GmbH
in Hamburg, Germany